IN

N ... AND

BIOMEDICAL

LABORATORIES

k due

SAFETY IN CLINICAL AND BIOMEDICAL LABORATORIES

Edited by
C.H. Collins

London
CHAPMAN AND HALL

First published in 1988 by
Chapman and Hall Ltd
11 New Fetter Lane, London EC4P 4EE
© 1988 Chapman and Hall

Printed in Great Britain at the
St. Edmundsbury Press, Bury St. Edmunds, Suffolk

ISBN 0 412 28370 0

British Library Cataloguing in Publication Data

Safety in clinical and biomedical laboratories
1. Medical laboratories—Safety measures
I. Collins, C.H.
616.07'5   RB36

ISBN 0-412-28370-0

CONTENTS

CONTRIBUTORS

C.H. Collins MBE, DSc, FRCPath, FIBiol.
Microbiology Department, Cardiothoracic Institute,
University of London, Fulham Road, London

W.J. Gunthorpe FIMLS, CBiol., MIBiol.
Central Health and Safety Unit, Inner London Education
Authority, London

D.A. Kennedy FIMLS, CBiol., MIBiol.
Supplies Technology Division, Department of Health and
Social Security, London

J.F. Stevens MPhil, MCB, CBiol., FIBiol., FRSH, FIMLS
St. Stephen's Hospital, London

A.E. Wright TD, MD, FRCPath, DPH, Dip. Bact.
Public Health Laboratory, Newcastle-upon-Tyne

PREFACE

Laboratory workers and managers in clinical and biomedical establishments who seek advice and immediate answers to questions on safe procedures may have to consult many different books and official publications. Parts of the 'Howie Code', for example, have been revised by no less than three additional publications in a few years. Books on chemical hazards abound but they are not geared to clinical and biomedical laboratories. Information on the hazards of equipment peculiar to those laboratories is widely scattered.

An excellent booklet, *Safety in Pathology Laboratories*, was published by the Departments of Health in 1972 but rapidly went out of print and was never revised. An attempt has been made, therefore, to fill these gaps and to collect, under one cover, information about the hazards, chemical, microbiological and equipment-related, that are not uncommon in these kinds of laboratories, and to give references to larger works and official regulations.

Information on health care of laboratory workers and the first aid procedures that should be applied in the kinds of accidents that may happen during their work are also included.

Finally, a check list is provided which may allow individual workers, safety staff and management to foresee and forestall hazards and accidents that might occur in their laboratories.

I am indebted to the contributors to this book, not only for their own chapters, but also for the material they supplied for the others.

C.H. Collins
Hadlow
June 1987

1 LABORATORY SAFETY – GENERAL CONSIDERATIONS

C.H. Collins

There is no hard evidence that clinical and biomedical laboratories are 'unsafe' places in which to work. Accidents do occur in them, of course, as they do in other workplaces. They contain fittings, equipment, chemicals and infectious agents all of which may, under certain circumstances, cause injury and illness. These may be harmless in themselves, but adverse contact with them may constitute a hazard. Before they are condemned out of hand however, they should be properly identified and the risk associated with them assessed in terms of what is reasonable and generally acceptable.

Assessment of risk is a quantitative and scientific procedure and must not be done on emotional grounds or for political reasons. The risk of working with, for example, a particular chemical, will depend on how it is contained or secured, the expertise of the people who handle it, and the precautions they take. Under some circumstances the risk may be minimal but under other conditions may be quite unacceptable. The most important aspects of risk assessment and of specifying precautions, therefore, are the identification and appreciation of the hazard and its potential for causing harm. Such information enables workers to decide what is safe, i.e. whether what they are doing is likely to cause injury or illness, and what they can do to avoid such harmful consequences to themselves and other people. Experience tells us that it is quite safe to work with

Escherichia coli on the bench in a clinical laboratory, but most unsafe to work with *Francisella tularensis*. Very few people have acquired infections while working with the former, but nearly everyone who has worked with the latter has become infected, no matter how much care was taken.

Safety in the laboratory is not just a matter for the worker. Other people have moral and legal obligations to ensure the safety of the worker, just as he or she has obligations to them. Moral obligations cannot be defined except in terms of philosophy and theology but legal obligations are imposed by most governments. In the UK it is the duty of the Health and Safety Commission and its Inspectorate to ensure that these obligations are fulfilled by both employers and employees.

In the following pages many of the hazards that exist in clinical and biomedical laboratories are identified. Attempts are made to assess the risks and advice is given on precautions that may be taken to ensure the health and safety of the laboratory workers and their colleagues and to satisfy the strict but reasonable requirements of the Health and Safety Inspectorate.

1.1 SAFETY POLICIES AND CODES OF PRACTICE

In some countries, including the UK, employers are required by law to prepare and publish a statement of their policy on the safety of people in their employment. In addition, they are required to furnish their employees with Codes of Practice which clearly indicate the procedures which must be followed to ensure their safety and health.

1.2 OFFICIAL CODES OF PRACTICE AND GUIDELINES

In the UK the only 'official' codes of practice and guidelines are related to microbiology. The *Code of Practice for the Prevention of Infection in Clinical Laboratories and Post-mortem Rooms* (DHSS, 1978), known as the 'Howie Code' (after the name of the chairman of the working party that formulated it) has been amended in part by the Advisory Committee on Dangerous

Pathogens (ACDP, 1984) and the Health Services Advisory Committee (HSAC, 1985, 1986) and guidelines for work with the virus (HIV) of AIDS have been published (ACDP, 1986) but there appear to be no official codes or guidelines for clinical chemistry, haematology, or histopathology. Useful advice and guidance in these fields has been given, however, by various professional bodies and individuals and these are referenced in the appropriate parts of this book.

1.3 SAFETY OFFICERS OR ADVISORS, SAFETY REPRESENTATIVES AND SAFETY COMMITTEES

There is no legal requirement for the appointment of Safety Officers or Advisors. Large organizations make such appointments in their own, as well as in their employee's interests. In hospital and biomedical laboratories, members of the professional staff are often appointed to supervise safety activities and training on a part-time basis. They may or may not receive specific training. Safety Representatives are members of trades unions, appointed by employers at the behest of employees who are union members. They are usually given basic training by their unions, but this is rarely at a scientific level. Safety Committees usually consist of representatives of management, employees, trades unions and the safety officer or representative. Their usefulness depends on the enthusiasm of the members.

Information about the duties and responsibilities of safety staff and the organization of safety committees may be obtained from the local office of the Health and Safety Executive or the appropriate trades unions.

1.4 PROTECTION OF THE LABORATORY WORKER: THE 'BARRIER SYSTEM'

The general principle of safety in clinical and biomedical laboratories involves the placing of barriers between the hazard and the worker and between both and the community. This is shown schematically in Fig. 1.1. The primary barriers are placed around the hazard, the secondary barriers are around the worker and the tertiary barriers around the laboratory itself. Any barriers may be breached however, by human error and mechanical faults. A safety net is therefore necessary and this is the emergency service, which cleans up spillages, makes equipment safe and renders first aid and medical treatment.

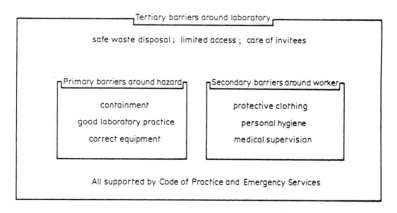

FIGURE 1.1 The barrier system for safety in laboratories.

1.5 PLACES AND PEOPLE

Before considering the hazards of the equipment and stock-in-trade of a laboratory it is desirable to look at the buildings themselves and the people who work in them.

1.5.1 The laboratory building and rooms

Very few laboratory workers have any voice in the design of the places in which they will spend most of their working lives. Very few laboratories have been designed by people who have actually worked in one. The outcome has been the unwitting introduction of hazards and of inconveniences that may father more hazards. Until recently there have been no legal standards or building regulations. Fortunately the inspectors of the Health and Safety Executive have applied the Howie Code and the recommendations of the ACDP to most, if not all, departments and sections of clinical and biomedical laboratories. This has raised the standards, particularly in hospital laboratories, even though these places have enjoyed 'Crown Immunity' until recently.

1.5.2 General design features

The following basic features are desirable and may be useful as a guide to laboratory workers who are fortunate enough to be involved in designing or refurbishing their laboratories.

1. At least 20 m³ floor space and 3 m bench run per person (some of both will be occupied by equipment).
2. There should be windows.
3. Ventilation, whether natural or mechanical should ensure at least six changes of air per hour.
4. Air should flow from corridors to laboratories and not in the reverse direction.
5. Heating should be under the control of the occupants.
6. Lighting should be arranged to give good overall illumination, casting no shadows on benches.
7. Each work place should have local lighting (e.g. angle-poise lamps).
8. Floors should be impervious to water and acid- and solvent-resistant and should be coved to the walls or to fixed furniture.
9. Bench surfaces should be fire-resistant, impervious to water and resistant to strong acids and alkalis, solvents and disinfectants.
10. Benches should be at a convenient height for work, depending on whether the staff will normally sit or stand.
11. Hand washing basins should be provided in each room.
12. Seating should be comfortable and designed to minimize backache.
13. A sufficient number of electrical, gas and water outlets should be provided to avoid trailing cables and tubing.

For information about the design of more specialized laboratories, e.g. those requiring microbiological safety cabinets, fume cupboards and large items of equipment readers are referred to Grover and Wallace (1979), Collins (1988), ACDP (1984), Everett (1985), Robb (1985), and current DHSS building notes on pathology laboratories.

1.6 SAFETY AWARENESS AND TRAINING

All members of the staff should be familiar with the local and general rules that are made for their safety. Each person should be given a copy of the local code of practice or guidelines, as well as copies of relevant national documents. Senior staff ought to set a good example of cleanliness, hygiene, tidiness and responsible behaviour. They have a duty to ensure that all other employees in their laboratories are adequately trained and know what they have to do, both in the ordinary way of work and in emergencies.

1.6.1 Training in safety

At present there are no formal courses in laboratory safety and little attention is given to the subject in colleges. Training schemes of various kinds, for use in colleges, laboratories and local classrooms, for all levels of laboratory staff are described by Collins (1988). Some attempt, at least, should be made to educate all laboratory employees at their place of work.

1.6.2 Professional staff

It should not be assumed that a professional qualification automatically indicates professional ability and a knowledge of safe procedures, even in the field in which the qualification was obtained. Trained chemists do not necessarily know how to deal safely with infectious materials; trained microbiologists are often ignorant of chemical hazards; workers in haematology and histopathology laboratories may be at a double disadvantage in that they are faced with both chemical and microbiological hazards as well as those peculiar to their trades.

TYPES OF MODERN FIRE EXTINGUISHERS

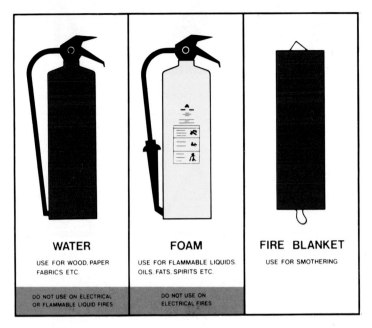

WATER

USE FOR WOOD. PAPER
FABRICS ETC.

DO NOT USE ON ELECTRICAL
OR FLAMMABLE LIQUID FIRES

FOAM

USE FOR FLAMMABLE LIQUIDS.
OILS. FATS. SPIRITS ETC.

DO NOT USE ON
ELECTRICAL FIRES

FIRE BLANKET

USE FOR SMOTHERING

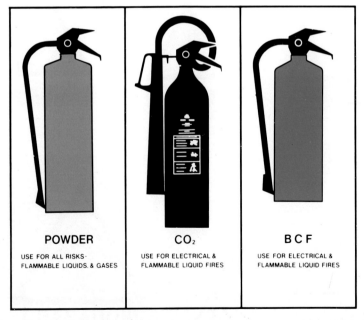

POWDER

USE FOR ALL RISKS·
FLAMMABLE LIQUIDS. & GASES

CO₂

USE FOR ELECTRICAL &
FLAMMABLE LIQUID FIRES

B C F

USE FOR ELECTRICAL &
FLAMMABLE LIQUID FIRES

The signs shown in these colour plates are by courtesy of Jencons (Scientific) Ltd., Leighton Buzzard, Beds. Full sizes and details can be found in their general catalogue no. 6, pages 245-271.

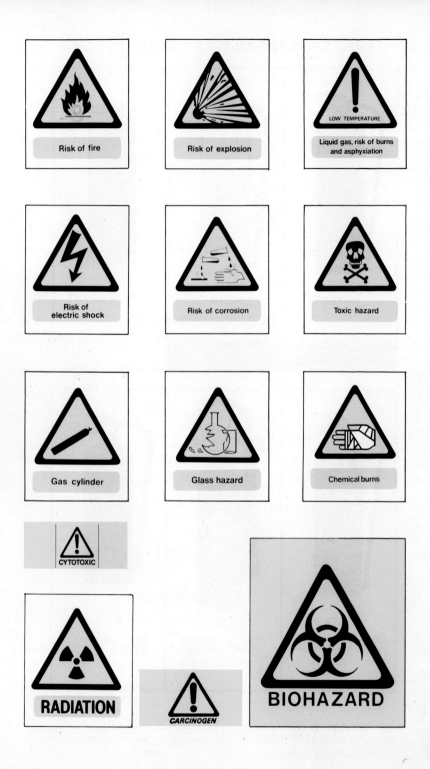

BURN WITHOUT OPENING

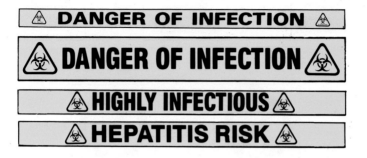

DANGER OF INFECTION

DANGER OF INFECTION

HIGHLY INFECTIOUS

HEPATITIS RISK

flammable

fire alarm

fire extinguisher

For use on electrical fires

For use on any fire

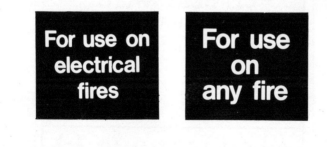

Water extinguisher

Dry powder

Foam extinguisher

Not to be used on electrical fires

CO₂

Eye protection must be worn

Laboratory coats must be worn

Wash hands

No mouth pipetting

Smoking, drinking and eating prohibited

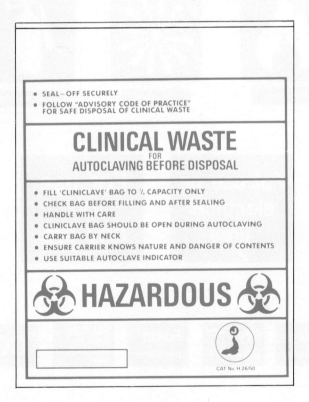

- SEAL—OFF SECURELY
- FOLLOW "ADVISORY CODE OF PRACTICE" FOR SAFE DISPOSAL OF CLINICAL WASTE

CLINICAL WASTE
FOR
AUTOCLAVING BEFORE DISPOSAL

- FILL 'CLINICLAVE' BAG TO ¼ CAPACITY ONLY
- CHECK BAG BEFORE FILLING AND AFTER SEALING
- HANDLE WITH CARE
- CLINICLAVE BAG SHOULD BE OPEN DURING AUTOCLAVING
- CARRY BAG BY NECK
- ENSURE CARRIER KNOWS NATURE AND DANGER OF CONTENTS
- USE SUITABLE AUTOCLAVE INDICATOR

HAZARDOUS

CAT No H 26/50

Professional staff change frequently. It is highly desirable therefore, that the attitude and knowledge of safe practices of new entrants (at any level) are assessed and are brought into line with local requirements.

1.6.3 Ancillary staff

Clerical, domestic and maintenance staff rarely understand the nature of laboratory work and may not appreciate the hazards inherent in certain materials, equipment, chemicals and cultures. Trained staff therefore have a duty towards ancillary staff. New recruits should have an induction course in which the procedures and potential hazards are explained and established staff should be reminded about the hazards regularly. The only officially published 'model rules' for ancillary staff are concerned with the prevention of laboratory acquired infections (see Chapter 6, Tables 6.5, 6.6 and 6.7).

Whenever possible, laboratory ancillary staff should be permanently allocated and not moved to other departments, thus reducing the need for repeated inductions and explanations.

1.6.4 Visitors

Visitors include staff of other departments, maintenance personnel (local and outside contractors), as well as visiting professionals and patients. The laboratory staff have a 'duty of care' towards such people and should escort them, generally supervise their activities and make sure that they do not come into contact with any hazardous or potentially hazardous equipment or materials.

1.7 PROTECTIVE AND OTHER CLOTHING

1.7.1 Laboratory overalls

The laboratory coat, originally designed by Dowsett and Heggie (1972) and its variations are now generally accepted as the most suitable protective garment for all kinds of clinical and biome-dical laboratory work. Some workers tend not to fasten the top section. This is not good practice. There is no reason to question the Howie Code which states that a sufficient supply of these overalls should be available to all workers, as and when required. No one should be allowed to work or walk about wearing a contaminated or dirty garment.

When not being worn these coats should be hung on pegs in the laboratory room in which the wearer works, not in corri-dors, staff changing rooms or other odd places. They should not be placed in the lockers that are provided for street clothing. Any contamination may not be obvious.

1.7.2 Aprons

It is recommended that plastic disposable aprons are worn over the Dowsett–Heggie garment for work with blood and Hazard Group 3 pathogens (see Section 6.9), and especially in inves-tigations with hepatitis and AIDS materials. While these aprons are suitable for work with some chemicals staff are advised to wear heavy-duty rubber aprons when dealing with strong acids and corrosive substances.

1.7.3 Gloves

The wearing of plastic disposable gloves is recommended for work with blood, for hepatitis and AIDS materials, and for some other pathogens. It is well known, however, that some manual sensitivity and dexterity may be lost by covering the fingers and common sense and experience should prevail. Heavy-duty rubber gloves should be worn for work with strong acids and caustic chemicals. Heat-resistant leather ('Pyro') gloves, *not* asbestos gloves, are necessary for handling hot materials, e.g. when unloading hot air ovens and autoclaves. The gauntlet varieties are safest. Cold-resistant ('Cryo') gloves should be provided in laboratories where solid carbon dioxide, liquid nitrogen and other cryogenic substances and materials are used.

1.7.4 Safety spectacles, goggles and vizors

In most industrial chemical laboratories the staff and visitors are required to wear safety spectacles. This might well be applied in clinical and biomedical laboratories where hazardous chemicals are used. Good safety spectacles fit over ordinary spectacles and are quite comfortable to wear. Goggles, which cover the whole of the area around the eyes, are preferable for manipulations with strong acids and caustic chemicals. Vizors present problems. Few people like wearing them and those that do not fit under the chin are not very good because splashes, steam and vapour still have access to the face. Vizors should be worn when autoclaves are unloaded – even when this is done properly (see Section 2.10.1.3).

1.7.5 Respirators

Full face respirators are rarely needed, but should be provided in large establishments, both for rescue work and entering rooms where infectious bacterial aerosols have been accidentally released. Training should be given in their use. Breathing equipment, i.e. with self-contained air supply, should not be available to any but staff specifically trained in its use.

1.7.6 Ordinary clothing and hair

There need be no restrictions on normal, and common-sense clothing, but accidents do happen to staff who are wearing very loose garments or clothing not under very good control.

Similarly, certain types of footwear may engender hazards. Flip-flops and open-toed sandals seem to attract broken glass. The former are frequently associated with tripping, which can be hazardous if the wearer is carrying chemicals or cultures. Accidents can happen when long, loose hair comes into contact with flames and moving parts of apparatus. Some restraint seems advisable.

1.8 SMOKING, EATING, DRINKING ETC.

No food or drink of any kind should be consumed or stored in the laboratory. Nor should it be stored in any laboratory refrigerator or cold room. On the other hand, refrigerators in staff rooms should be clearly marked so that no infectious materials or chemicals are placed in them.

Smoking should also be banned in laboratory rooms as part of the general principle that nothing should be placed in the mouth. By the same token, labels should not be licked and pencils and pens should not be chewed. The application of cosmetics to the face is also a hazardous activity in laboratory rooms, even if the hands have been washed first. Hand creams are permissible.

1.9 REFERENCES

Advisory Committee on Dangerous Pathogens. (1984) *Categorisation of Pathogens According to Hazard and Categories of Containment*. London: HMSO.

Advisory Committee on Dangerous Pathogens. (1986) *LAV/HTLVIII – the Causative Agent of AIDS and Related Conditions – Revised Guidelines*. London: Health and Safety Executive.

Collins, C.H. (1988) *Laboratory Acquired Infections*. 2nd edn. London: Butterworths pp. 203–221.

DHSS (1978) *Code of Practice for the Prevention of Infection in Clinical Laboratories and Post-mortem Rooms*. London: HMSO.

Dowsett, E.G. and Heggie, J.F. (1972) A protective laboratory coat. *Lancet*, **i**, 1271.

Everett, K. (1985) Planning biological laboratories. In *Safety in Biological Laboratories* (ed. C.H. Collins). Chichester: Wiley. pp. 67–74.

Grover, F. and Wallace, P. (1979) *Laboratory Organisation and Management*. London: Butterworths. pp. 1–32.

HSAC (1985) *Safety in Health Service Laboratories: Hepatitis B*. Health Services Advisory Committee. London: HMSO.

HSAC (1986) *Safety in Health Service Laboratories: The Labelling, Transport and Reception of Specimens*. Health Services Advisory Committee. London: HMSO.

Robb, J.R. (1985) Organising the design of a safe laboratory. In *Handbook of Laboratory Health and Safety Measures*. Lancaster: MTP Press. pp. 1–11.

2 EQUIPMENT-RELATED HAZARDS

D.A. Kennedy

Some of the hazards that are encountered in clinical and biomedical laboratories result from design and manufacturing errors in equipment and materials. Others are the consequence of misuse of equipment which is inherently safe.

In the United Kingdom the Supplies Technology Division (STD) of the Procurement Directorate of the Department of Health and Social Security (DHSS) is responsible, among other things, for the investigation of accidents with, and reports of, defects in all kinds of equipment and materials used in the National Health Service (NHS). For many years senior NHS staff have been encouraged to ensure that accidents and serious defects are reported because it is recognized that these often have a significance beyond the hospital in which they are detected, and sometimes beyond the NHS.

A small group within the STD is responsible for equipment and consumables (e.g. reagents, diagnostic kits and calibrants), that are used in pathology departments. This group investigates defects and accidents in order to take remedial measures and to prevent recurrence. As a rule, manufacturers readily involve themselves in these investigations because they recognize the opportunity to improve the safety and performance of their products.

Typically, an investigation includes a visit to the site and discussion with the laboratory staff and the manufacturers' representatives. Sometimes detailed tests have to be carried out to help determine the cause of an accident or to prove a safety modification. If necessary, the DHSS can issue a warning to the NHS. This can take the form of a 'Hazard Notice', if, for example, immediate action is necessary because there is a serious risk to users. In less urgent cases it may be presented as an article in the next issue of the (DHSS) *Safety Information Bulletin*. Before a warning notice is issued the manufacturer is invited to comment on the draft. When wider issues are involved trades associations and professional bodies involving NHS staff are consulted.

Much of the material in this chapter was derived from investigations by the STD group responsible for pathology laboratory equipment and consumables.

2.1 CLASSIFICATION OF EQUIPMENT-RELATED HAZARDS

There are several methods of classifying equipment-related hazards. One method considers the potential of the equipment or system (where there is combination of equipment and chemicals, as in continuous-flow chemical analysis) to cause injury to the user by the direct transfer of some form of energy. This epidemiological approach was elaborated by Haddon *et al.* (1964). A good example is electrically unsafe equipment which gives a shock to the user.

Steere (1980) classified the types of energy, or hazards, that can cause damage directly or indirectly. He included electrical and mechanical hazards in the class that have direct action, and gave chemicals as examples of a class in which there is a dual potential. Sulphuric acid, for example, used as an analytical reagent, may cause immediate injury by burning, but carbon monoxide is harmful indirectly by combining with haemoglobin and limiting oxygen transport to tissues. Biological agents and atmospheric pressure differentials are classified among those having an indirect action: both have a long term effect on the human body.

A third method is to classify equipment-related hazards according to the causes of specific accidents. In the experience of the STD they can be roughly classified as follows (Kennedy, 1979a):

1. Those involving equipment intended for laboratory use and in which the construction is faulty.
2. Those due to improper (i.e. careless or negligent) use of equipment intended for laboratory use; or improper adaptation of equipment not originally designed for laboratory use.
3. Those attributable to lack of proper maintenance whether it be the responsibility of the user or the manufacturer's specialist staff.

This retrospective and deterministic approach has stood the test of time and has been justified by results in the prevention, or at least a reduction in the recurrence of incidents. It is important, however, that all three methods are used when the equipment is part of a system that includes the user, other people, and possibly other equipment. It is then convenient to view an accident that involves equipment as a system which can be broken down into its component parts (agent, cause and effect), to arrive at the best course of action to prevent recurrence. Furthermore the process may be made prospective by considering all three classification methods in the evaluation of the hazard content of new equipment and systems. Thus the finding that a user has received an electric shock from a spectrophotometer because he spilled conducting fluid on it may be seen as an electric shock hazard, a hazard due to improper use, and a hazard peculiar to spectrophotometers. Here, there are remedial lessons for both user and manufacturer.

In the outline case studies presented later in this chapter some will be grouped according to the agent, some according to Kennedy (1979a) and others according to the type of equipment.

2.2 SOME AGENTS INVOLVED IN EQUIPMENT-RELATED HAZARDS

2.2.1 Fire

Fire has a great potential for causing injury and loss of life through burning and through asphyxiation caused by the inhalation of smoke. It is also responsible for considerable financial loss. Fire can spread rapidly from the source, e.g. a laboratory, threatening not only the site and its contents, but also the entire building, neighbouring buildings and their occupants. Hospital laboratories usually have all the materials necessary to start and propagate a fire: sources of ignition, fuel and oxidants. Histology laboratories offer the most hazards in this respect and Steere (1980) mentions a fire in a tissue processor that necessitated the evacuation of patients from a nine-storey hospital and cost US$1 million. This fire was probably caused by an electrical fault.

Kennedy (1985) has discussed 21 incidents, fires and over-heating episodes in histological equipment used for paraffin wax embedding techniques. In addition the *DHSS Health Technical Memorandum No 83* (DHSS, 1982b) gives advice on fire precautions in health care premises including laboratories.

2.2.2 Electricity

Some electrical incident statistics involving laboratory equipment are shown in Table 2.1.

TABLE 2.1 Some electrical laboratory equipment incidents reported to the STD* July 1973–November 1985

Total number of reports followed up	165
Electric shock	26
Fire, overheating or explosion	40
Serious design or constructional deficiencies with potential risk for shock or fire	65
Others	34
Hazard Notices issued	21
Lesser warnings, including notes in Safety Information Bulletin	19
Remedial action by manufacturer without DHSS action	41

* Supplies Technology Division of the Department of Health and Social Security

Fires arise in electrically operated equipment because of overloading of components and short circuiting. The hazard is exacerbated by the presence of flammable material inside the equipment and by spillage and leakage of conducting liquids. An electric shock may be nothing more than a mild tingling sensation but can be a more painful stimulus, resulting in total loss of muscle control and death. Large electrical currents may cause severe burns. The quantitative effects of an electric current at 60 Hz are given by Steere (1980) and these are of value when considering the effects at 50 Hz (used in the UK). More details are given by the *International Electrotechnical Commission* (IEC, 1984). Electric shocks, however mild, can also generate other hazards by causing a reflex – the 'startle reaction' – that may cause the victim to lose control of a chemical or biological substance with which he is working. For example, some bacteriologists received quite minor shocks from a loop-sterilizing device but as some of them were working with tuberculous material the results could indirectly, have been more serious (personal communication).

Electrical sparks are another problem. They may ignite flamable vapours, e.g. in a refrigerator, as described later. It is DHSS policy that all NHS laboratory equipment should comply with the *Electrical Safety Code for Hospital Laboratory Equipment* (ESCHLE; DHSS, 1986). Furthermore, NHS purchasers are advised to follow the DHSS code of practice for acceptance testing of electrically operated hospital laboratory equipment (DHSS, 1986). This code is intended to establish adequate safety standards in the design, and construction of laboratory equipment for use in hospitals and other health care establishments with the object of ensuring:

1. Prevention of electric shock to operators.
2. Prevention of overheating and risk of fire in both the equipment and its surroundings.
3. Prevention of mechanical instability and contact with moving parts.
4. Provision of adequate documentation and clearly marked controls.

2.3 EQUIPMENT-RELATED EXPLOSIONS, BURSTING INCIDENTS AND SIMILAR PROBLEMS

These have the potential to cause injury, perhaps to many people.

2.3.1 Azides

Twelve chemical explosions have been attributed to the discharge into copper waste pipes of effluents containing azides from blood cell counters and other equipment (see Table 2.2). Freeze-drying systems used to process azide-containing materials may discharge hydrogen azide into the surrounding atmosphere. This substance is extremely toxic (a Threshold Limit Value of 0.1% would be appropriate). A Hazard Notice (DHSS, 1982) gives details of this hazard and of other Warning Notices about the use of azides.

TABLE 2.2 Azide-related explosions in pipework (Data collected by the STD*)

---

1. There was a small explosion in the copper U-bend of a sink when an attempt was made to clear a blockage with a screwdriver. The sink received azide-containing effluent from a cell-washing centrifuge.
2. Later, there was a more violent explosion below the same sink and under similar circumstances. The plumber required medical attention to eyes and hand.
3. A copper pipe that carried azide effluent from a blood grouping machine exploded when attempts were made to unscrew a joint. Smaller explosions occurred when attempts were made to cut the pipe. Some of the contents of the pipe spilled on the floor and dried. There was a small explosion when a plumber walked on the dried material.
4. A violent explosion occurred in a copper waste trap when a cupboard door, approximately 6 m away, was slammed. The bottom of the trap was blown open and fragments of copper were blown away. The trap was part of a system used to dispose of effluent from a blood cell counter.
5. A small explosion occurred during an attempt to clear a blocked pipe that carried effluent from a blood cell counter.

continued

TABLE 2.2 *continued*

6. There was another explosion when an attempt was made to clear a blocked pipe with a 'sanisnake'. This pipe had also carried effluent from a blood cell counter.

7. There was an explosion in under-bench plumbing that had been used over a long period to dispose of sodium azide (calculated amount 22 kg) contained in the effluent from a blood cell counter. One worker was struck by a piece of metal that had ricocheted from the wall and sustained laceration on the lateral chest wall. Another had a small middle ear haemorrhage as a result of the blast.

8. Minor explosions occurred when pipework, used to dispose of blood cell counter effluent, was decontaminated by heating with a welding torch. There was another explosion when another length of this pipe was being decontaminated by heat. The metal was propelled approximately 8 m and the fitter's right hand was lacerated.

9. There was an explosion when attempts were made to unsolder a joint in a copper pipe through which azide effluent was discharged.

10. Another explosion occurred in a length of copper pipe carrying effluent from a blood cell counter. The pipe had been painted white and it was assumed that it was made of plastic.

11. There was an explosion in a nickel-plated tube that was part of a suction jar assembly used in manual cell washing. The saline used contained 0.1% sodium azide. The detonation occurred while a cleaner was emptying the jar. She allowed the tube to strike the lip of the jar which was shattered by the explosion. Fortunately she suffered only temporary deafness.

12. An explosion occurred in copper pipework which was part of a commercial freeze-drying plant used to process animal sera containing sodium azide.

NB No blood cell counting reagents currently used in the UK are known to contain azide.

* Supplies and Technology Division of the Department of Health and Social Security.

2.3.2 Perchloric acid

The hazards of perchloric acid are well documented by Everett and Graf (1971). It was responsible for a number of incidents in the NHS over ten years ago. A continuous flow method for the determination of protein-bound iodine used 70% perchloric acid and concentrated nitric acid in the wet ashing of samples.

In one incident the effluent, which also contained salts of arsenic, was discharged into an iron waste pipe that ran down the side of the building housing the laboratory. This pipe was damaged and the effluent gushed out, contaminating the surrounding brickwork and also the the woodwork of an adjacent cedar-clad structure. Even the carpet in an office on the floor below the laboratory gave a positive reaction to tests for perchlorate. The removal of contaminated structures and material was expensive and hazardous. In another incident some 2.5 l of perchloric acid were spilled on a wood block floor which was not effectively decontaminated. About three years later it ignited and a member of the staff suffered burns.

Perchloric acid is no longer employed in the estimation of protein-bound iodine, but this hazard is included here in case its use is contemplated for any other purpose.

2.3.3 Picric acid

Long-term use in a hospital laboratory of a continuous-flow biochemical analyser for the estimation of creatinine by the Jaffe reaction led to gross contamination of laboratory benches, floors and wastepipes. Picric acid and its salts, especially those of lead, calcium, iron and copper, can be highly explosive when dry. Picrates are not spontaneously explosive but are readily detonated by impact or heat. In this incident a lead waste pipe that received effluent from the analyser had a long-standing defect with the result that the concrete floor of the laboratory basement was considered by explosive experts to be in a dangerous condition because of the presence of picrates. The laboratory was obliged to close by the Health and Safety Executive pending an expensive decontamination exercise by specialist contractors.

2.3.4 Bursting incidents

Some bursting incidents have resulted from the use of domestic pressure cookers and microwave ovens to heat bottled culture media (MRC, 1985). Large pressure vessels have been involved in more serious incidents (see section 2.10.1).

2.4 MECHANICAL HAZARDS

This is a diverse group in which the hazard content is the result of contact between a person, or part of him, and an object when one or both are moving. Steere (1980) outlines the hazards of rotating machinery in the laboratory and states that the most common mechanical hazards are due to exposed V-belts and pulleys on vacuum pumps. So far no incident involving these has been reported to STD and in our experience such pumps used in the NHS are fitted with adequate guards as advocated by Steere.

Other examples include centrifuges (considered below) and injuries caused by sharp or pointed objects (e.g. a cut finger caused by the sharp edge of a metal flammable liquid storage cabinet - an incident recently investigated by STD). Failure of the spring-loaded hinges of a cryostat microtome and the sudden collapse of the support of a flame photometer while in use, are good examples of mechanical hazards caused by poor design.

2.5 COMPRESSED GAS CYLINDERS

Compressed gas cylinders, which should be secured upright with chains or straps, offer hazards if they are moved without care (see also Section 4.1.2). Johansen (1984) recounts an incident where a gas cylinder was being moved by sliding it across the floor with the valve end propped on the shoulder of the worker. Suddenly the valve came off and he found himself wrestling with an uncontrollable 110 kg steel monster. The cylinder got out of control and careered around the room. A painter was knocked off his scaffold, concrete blocks were broken away from a wall and an electrician had to run for his life. Another cylinder was tilted and its valve bent, but fortunately not broken. Purpose-built trollies should always be used to move compressed gas cylinders.

2.6 GLASS HAZARDS

Cuts from broken glass are very common accidents in laboratories (Ederer *et al.*, 1971; Hellings, 1981; Kibblewhite, 1984; Hamilton, 1984). Perhaps the most common in clinical laboratories are punctures of the skin, usually of the hand by glass Pasteur pipettes and by broken or chipped graduated pipettes, especially when the latter are incorrectly forced into pipetting devices.

Glass desiccators may implode. The STD investigated one such incident where the strength of the desiccator was inadequate and it was not inside a protective cage, as is generally recommended. The STD is currently investigating an incident in which the user of an amino acid analyser received a serious cut to the hand from the fragments of a plain glass bottle which burst while in use as a reagent reservoir and which was pressurized to 1 atm. The manufacturer had recommended that a plastic-coated bottle be used but a clear glass one was substituted because it was easier to see any turbidity due to bacterial growth.

Willhoft (1986) has reviewed the hazards of carbonated soft drinks packaged in glass bottles. In particular he emphasized the hazard resulting from a large headspace above the liquid. This was the case with the amino acid bottle incident mentioned above and has some relevance in clinical laboratories.

The deliberate breaking of glass, e.g. for making glass knives for ultramicrotomy, is hazardous. Much care is needed.

2.7 IONIZING AND NON-IONIZING RADIATIONS; HEAT

In the UK the use of ionizing radiation is governed by the *Ionizing Radiations Regulations, 1985*. There is also an *Approved Code of Practice* (HSE 1985) for the protection of persons against ionizing radiations arising from any work activity.

2.7.1 Electron-capture detectors

These incorporate sources of beta-radiation and are used in gas chromatography systems. No hazards involving these have been reported to the STD.

2.7.2 X-rays

Weeks (1976) and Edmunds and Sutton (1985) report cases where users of X-ray diffraction spectrometers received radiation burns on the hand. In the case described by Weeks there was evidence that the victim had made adjustments with the beam switched on; in that described by Edmunds and Sutton the victim had removed a shutter plate and exposed his hand to the beam. Weeks reviews other cases and lists the many warnings that have been given about this kind of X-ray producing equipment. In particular he states that no X-ray producing equipment should be used unless an interlock is installed in such a way that access to the beam is impossible while the machine is in operation. Such interlocking is a requirement of the *Approved Code of Practice* (HSE, 1985)

2.7.3 Ultraviolet radiation

Germicidal ultraviolet (uv) sources, e.g. at 100–280 nm, have been installed in some microbiological safety cabinets and culture media plate pouring machines. In one recent case the worker was exposed to uv while sitting at a Class II safety cabinet and in another two users were exposed to uv after they had removed a cover from a plate-pouring machine to deal with frequent blockages caused by petri dishes that jammed the conveyor. These people later suffered from the classical 'sand in the eyes' symptom of conjunctivitis, as well as erythema around the eyelids.

2.7.4 Lasers

If shone into the eyes, lasers can cause eye lesions, depending on the power density, wavelength, duration of exposure and size of the retinal image. Lasers are used in some blood cell counters and cell-sorting instruments but there is no evidence that they present a hazard to the operators providing that they are used in accordance with the manufacturer's instructions. British Standard 4803 (BSI, 1983) gives guidance on the protection of users of laser radiation. DHSS (1984c) gives guidance on the safe use of lasers in medical practice.

2.7.5 Ultrasonic equipment

Ultrasonic cleaning baths and cell disruptors, both of which operate at cavitational frequencies, are often used in biomedical laboratories. Williams (1985) points out that in these baths the transient cavitation gives rise to a substantial amount of white noise at the first subharmonic of the driving frequency. This intense subharmonic signal is usually within or close to the audible range and consequently poses a significant threat to hearing unless the bath is· placed within a soundproofed cabinet or some kind of ear protection is worn. It is prudent to adopt the same precautions with ultrasonic cell disintegrators. Some bench-top equipment available in the UK is provided with soundproofing by the manufacturers but other equipment is designed to be held in the hand and with this ear protection is advisable.

Williams (1985) warns that users of ultrasonic cleaning baths may suffer petechial lesions if the bare hands are immersed. This may engender allergic dermatitis if the exposure enables a detergent or caustic cleaning material to penetrate the skin. He recommends that rubber gloves are worn if the hands are likely to be immersed while the bath is in use. He also warns that these baths may generate aerosols from the fluids in the bath. This is also a possible hazard when cell disrupters are used with pathogenic organisms. These aerosols may cause damage or infection of the respiratory tract. Ultrasonic cleaning baths should be well ventilated, preferably at a negative pressure, e.g. in a fume cupboard with the access sash open.

2.7.6 Visual display units (VDUs)

VDUs are now widely used in clinical and biomedical laboratories and some concern has been expressed about the possible health effects of the radiation emitted, especially on pregnant women. Current opinion is that these levels are very low and that any discomfort suffered by users may be due (among other things) to poor siting, which leads to physical strain, and to poor lighting, which affects the eyes. The Health and Safety Executive (HSE, 1986) gives some guidance on working with VDUs and has also published a bibliography on the health effects of VDUs (HSE, 1984). The topic has aroused much interest among trades unions and several reviews have been published (TUC, 1985; APEX, 1985; ILO, 1986).

2.7.7 Heat

Hot surfaces that are not seen to be hot may offer a hazard to the user. The *Electrical Safety Code for Hospital Laboratory Equipment* (ESCHLE; DHSS, 1986) gives a list of temperature values that should not be exceeded in normal conditions, in accessible parts of equipment. For example, the temperature of metal handles should not exceed 55°C and accessible parts that may be touched inadvertently should not exceed 100°C.

Large autoclaves and large biochemical analysers which incorporate flame photometers and some heating baths may radiate so much heat that the room temperature is raised considerably. While heat stress is unlikely, working conditions may become uncomfortable and air conditioning desirable. Gill (1980) gives excellent coverage of the occupational hygiene aspects of heat. Moist heat from steam and the escape of hot liquids can cause serious injuries. HSE (1976) outlines an incident, all too common in the early days of bacteriology, in which a technician suffered extensive scalding when bottles of culture media burst because they had not been allowed to cool before they were moved.

Burns from microwave ovens, including those due to direct exposure because the source was not automatically switched off when the door was opened are reviewed by Maley (1986) but no laboratory accidents are mentioned.

2.8 MISCELLANEOUS AGENTS

Under this heading we may place the ergonomic shortcomings of equipment that could lead to fatigue and perhaps back strain. These include stress-inducing factors such as noise and unreliability, badly designed furniture and absence of lifting and moving equipment. Pearce and Shackel (1979) review the ergonomics of scientific instrument design and advocate that the designer should consider the user. Matteson and Ivancevitch (1982) investigated the sources of job-related stress and associated factors in a sample of American medical technologists. Overall, they found that while 'pressure for immediate results' was the most frequent source of stress and was ranked one on a scale of one to 22 in a sample of 462 questionnaire returns, equipment failures were ranked at six. It is easy to speculate that the working lives of those who suffer from pressure for immediate results are not likely to be eased by equipment failure! Minok *et al.* (1982) tell of a laboratory worker who developed mild osteoarthritis in the interphalangeal joint of her right thumb and this was attributed to repetitive pipetting. They record that she performed serological tests for hepatitis and during one of her typical activities was required to exercise the affected joint about 20 000 times.

Back problems, which may be caused by uncomfortable chairs and benches at the wrong height, as well as through moving and lifting heavy loads, are common occupational disabilities. Gas cylinders are among the heaviest articles that laboratory workers are required to move and as noted above, trollies should be provided for these and for any other heavy loads.

2.9 EXAMPLES OF HAZARDS CLASSIFIED ACCORDING TO KENNEDY (1979a)

2.9.1 Faulty design or construction of laboratory equipment

Fires in electrically-heated equipment (Kennedy, 1985 and Section 2.2.1) and the history of electrical problems in general (Table 2.1) illustrate this class of equipment-related hazard. The DHSS was acutely aware of the need to develop and maintain safety standards for equipment. Continual electrical defects stimulated it to re-examine the original version of ESCHLE to see if it could be made more precise in its requirements, clearer in its meanings, more up to date and, most important, whether a full range of laboratory equipment that complied with its requirements could be placed on the market.

A programme of test work was agreed in cooperation with the British Laboratory Ware Association, a leading trade organization. The initial testing programme was funded by the DHSS and carried out in the Testing Service Laboratories of the British Standards Institution (BSI). A number of items was examined, as was the usefulness of ESCHLE itself. A group of experts, including those from the trade, discussed the changes that were required to answer criticisms that had been made of ESCHLE (which was first published in 1979). As a result the standard was made workable and suppliers began to obtain BSI Certificates of Compliance.

Wax embedding centres were among the first items to be examined and observations were published in *Health Equipment Information* (HEI; DHSS, 1984a). Since then details of other certificated equipment have been published in that journal. NHS purchasers are encouraged by the DHSS to purchase only equipment that complies with ESCHLE, preferably that which is certificated as such, if it is available.

2.9.2 Improper use and adaptation of equipment

2.9.2.1 *Anaerobic equipment* A good example of improper use is illustrated by an explosion in an anaerobic chamber in

use in a microbiological laboratory. The user had attached a cylinder containing a mixture of 90% hydrogen and 10% carbon dioxide instead of the gas mixture recommended by the manufacturer, which contains 80% nitrogen, 10% carbon dioxide and 10% hydrogen. There was a violent explosion which blew the heavy cabinet off the bench and wrecked it. After this, and a second, less dramatic explosion, some experiments were carried out to determine if the gas mixtures recommended by manufacturers were safe to use under normal conditions of temperature and pressure. The hitherto safe operating record of anaerobic cabinets, now generally in use, was also reviewed. The conclusions were that providing the manufacturers' instructions were followed there was no danger of explosion with the following mixtures (v/v%):

	CO_2	H_2	N_2
1.	5	10	85
2.	10	10	80
3.	10	5	85

To distinguish these mixtures from those that are unsuitable, and which, in the UK, have identical colour codes (BSI, 1973a 1973b) the British Oxygen Company agreed to mark cylinders containing the approved mixtures with the words 'ANAEROBIC GROWTH MIXTURE' painted vertically down the body, and to indicate the composition of the mixture on the shoulder of the cylinder.

Explosions have occurred in anaerobic jars that use the evacuation-displacement principle and in those that use commercial gas-generation sachets. Experiments have shown that in the evacuation-displacement method it is possible to achieve an explosive mixture (between 4% and 40% hydrogen in air) and with gas-generation sachets an explosive mixture is inevitable. If loose pellets or fragments of the catalyst are present or exposed these can become hot enough to ignite the explosive mixture, as can the metal gauze catalyst envelope if it contains insufficient catalyst. Users should note that hydrogen requires only one tenth as much energy for ignition as other fuels (British Cryogenics Council, 1970).

Gas mixtures containing more than 10% hydrogen are not recommended for anaerobic jars.

In the absence of specific instructions by the manufacturers on the use of gas-generation sachets in anaerobic or catalyst jars the following precautions should be taken:

1. At least 1 g of catalyst should be used per litre volume of jar.
2. The catalyst should be renewed at least once each month.

More information on the use and care of anaerobic jars is given by the DHSS (1979).

2.9.2.2 Gas cylinder reducing valves A good example here is of a fire that occurred in an oxygen cylinder reducing valve which had been in use for 7–8 years without any maintenance. It was fitted to a cylinder that supplied oxygen to a flame photometer burning a mixture of oxygen and propane. After a period of uneventful use a long, laser-like flame was seen issuing from the valve. The operator immediately turned off the oxygen with the cylinder key, thus preventing a serious fire. Examination of the valve revealed a hole that had burned through the diaphragm chamber and long-standing damage to the sintered nickel filter inside the stem. It was likely that nickel particles had broken free from the damaged filter and been swept by the gas stream into the diaphragm chamber where they had been spun round at a very high speed. The friction generated heat which ignited the gas. This burned a hole through the chamber and the flame escaped into the laboratory. The manufacturer of the valve, who had not supplied it directly, commented that the sintered nickel filter should have been changed annually. This important information had not been passed on to the user.

Gas cylinder reducing valves require expert maintenance in accordance with the manufacturers' specifications. Correct storage of cylinder valves is essential and guidance is available (DHSS, 1984b).

2.9.2.3 Flame photometers A few years ago there were a number of fires in flame photometers of the type associated with the incident described above. The burner assembly of these instruments had a metal capillary, used to inject diluted plasma into the flame. This screwed into the body of the burner assembly. The capillary tended to become blocked, when it was necessary to disconnect the gas hoses, remove the assembly and dismantle it to remove the blockage. Regular cleaning of the capillary was recommended by the manufacturer as a preventative measure. The fires occurred when lighting up after this procedure. Investigations showed that these were due to a gas leakage caused by failure to replace a PTFE washer which was fitted to ensure a gas-tight joint. In some cases a damaged washer had been put back. In another incident it appeared that the gas hoses, fitted to the burner assembly, had not been properly tightened.

As this instrument appeared to lend itself to improper maintenance, and other models, not requiring oxygen became available, it was recommended that it be phased out of service in the NHS.

2.9.3 Improper use of equipment not intended for laboratory use

2.9.3.1 *Refrigerators*   Kennedy (1979a) describes an explosion in a histology laboratory caused by the storage of 2.5 litre bottles of diethyl ether, which had leaking closures, in a domestic-type refrigerator. This appliance was not spark proofed and the contact breaker points of the thermostat were, as is usual, sparking continually. The explosion, which was inevitable, wrecked the refrigerator and badly damaged the laboratory, causing large cracks in the walls and ceilings. Fortunately there was no fire and nobody was in the laboratory at the time.

There was an explosion in a refrigerator in which a chromatography plate, containing traces of a flammable solvent, was stored during a lunch break, and in another where laboratory rats, killed by exposure to diethyl ether and awaiting autopsy, were stored. In other laboratories spark ignition of flammable solvents has started serious fires. Steere (1980) states that if opened containers of ether, isopentane and similar solvents with high vapour pressures and very low flash points (i.e.

below the temperature of the appliance) are stored in refrigerators or deep freezers violent explosions may be expected.

If it is really necessary to store flammable liquids in refrigerators then they should be in small amounts, in sound containers with leak-proof closures, preferably packaged inside larger containers. All sources of ignition, e.g. thermostat contacts and light switch mechanisms, should be outside the refrigeration chamber.

Intrinsically safe refrigerator cabinets are available commercially but they are expensive. The Common Services Agency of the UK Department of the Environment (DOE) has published specifications for refrigerated storage cabinets for flammable substances (DOE, 1980). Refrigerators that are not intended for the storage of flammable substances should be labelled as such. Labels are available commercially (see colour plate section).

2.9.3.2 Vacuum flasks A vacuum flask of the type used for picnics etc. was used to transport liquid nitrogen. As the nitrogen was poured into the flask some leaked into the cavity between the silvered glass and the plastic outer case. When the nitrogen reverted to its gaseous phase the increase in volume could not be contained and the vessel exploded. Fortunately no one was injured. Only specially designed Dewar vessels should be used for the transportation and storage of liquid gases.

2.10 HAZARDS OF PARTICULAR TYPES OF EQUIPMENT

British Standard 2646 (BSI, 1955) is a specification for autoclaves used for the sterilization of bacteriological culture media and similar purposes. British Standard 3970, Parts 1–5 (BSI, 1966) is a series of specifications concerned with steam sterilizers (i.e. autoclaves). *Health Technical Memorandum 10* (DHSS, 1980) gives advice on the choice, purchase, installation, testing and commissioning, use and maintenance of steam sterilizers used for different purposes including bottled fluids.

2.10.1 Autoclaves and other pressure vessels

The Public Health Laboratory Service (PHLS, 1978) gives advice on the safe and efficient operation of autoclaves. Three main hazards are identified.

1. Failure to sterilize, i.e to make safe infected materials.
2. The inherent pressure vessel hazard.
3. The unloading hazard.

2.10.1.1 *Failure to sterilize*   This is important because infected material may not be made safe and staff handling it after ineffective autoclaving are at risk from pathogenic organisms. It is known that inadequate venting of the chamber of an autoclave controlled by pressure is a significant cause of failure to sterilize. Collins (1988), however, searched the literature and did not find a single case of laboratory-acquired infection due to the failure of an autoclave to decontaminate an infectious load. He reports finding unsuitable autoclaves in some new microbiology laboratories and argues that the more sophisticated and complicated an autoclave the more likely it is to go wrong, be misused or misunderstood by those who work in laboratory wash-up rooms.

2.10.1.2 *The pressure vessel hazard*   In the experience of the STD even simple equipment may be misused. We have investigated an incident in which a domestic type pressure cooker was destroyed by an exploding bottle. The conclusions are that with these and any small laboratory autoclaves the precautions shown in Table 2.3 should be taken.

TABLE 2.3 Precautions with pressure cookers and small autoclaves

1. All bottle caps should be loosened.
2. An adequate volume of water should be placed in the vessel each time it is used.
3. The vessel should not be allowed to boil dry or be left unattended whilst in use.
4. The manufacturer's instructions should be observed.
5. The vessel should be inspected daily for signs of corrosion and the vent inspected daily for evidence of blockage.
6. Only replacement parts supplied by the manufacturer should be used.

All larger and conventional laboratory autoclaves should now be fitted with an interlocking safety device on the door to prevent opening until the pressure has returned to atmospheric and the temperature fallen to 80°C or less within a sealed container. Advice on coping with the pressure hazards of larger and conventional laboratory autoclaves is given by the PHLS (1978) and is reproduced in Table 2.4.

TABLE 2.4 Precautions to be taken when loading laboratory autoclaves

1. Inspect for any visible damage to the door seal.
2. Check the load to ensure that all caps are loose on screw-capped bottles and that there is a generous headspace (not less than 1/3) above the liquid in each bottle. Remove any cracked or otherwise defective bottles.
3. If the inclusion of sealed containers is unavoidable or if the load is unusually large or consists of large unit volumes, display a notice that special care is necessary on this account.
4. Close the door in the approved manner (if necessary referring to the written operating procedure, which should be displayed in the same room) and do not misuse the closing gear.
5. Check that the door is properly closed.
6. Make no attempt to override any safety mechanism.

At the end of the run and *before* attempting to open the autoclave door:

1. Ensure that the main steam inlet valve is closed (or its equivalent if steam has to be generated on the site).
2. Open the air break valve and the exhaust valve.

From The Public Health Laboratory Service Subcommittee on Laboratory Autoclaves Report (1978). Autoclaving practice in microbiology laboratories. *Journal of Clinical Pathology,* **31,** 418–22. Reproduced by permission of the editors of the *Journal of Clinical Pathology.*

2.10.1.3 *The unloading hazard* Reference has already been made to the bursting of bottles after they have been removed from an autoclave before they have been allowed to cool down. HSE (1976) records such a case where the operator was injured, but found that although he had been issued with full protective clothing he was not wearing it. Furthermore, he did not follow the instructions about the cooling down period because he was in a hurry to go home. It was considered that a greater degree of safety could have been provided by fitting a temperature-sensing device in the autoclave to prevent the opening of the door before the contents had cooled to a safe temperature. Advice on dealing with unloading hazards are given by the PHLS (1978) and are reproduced in Table 2.5.

TABLE 2.5 Precautions to be taken when unloading laboratory autoclaves

---

1. Wear a safety visor when opening the door (if possible, the door should be only 'cracked' open (½ to 1½ in.; 12 to 36 mm) and the autoclave left for 15 min to accelerate the further cooling of the load before the actual unloading). During these operations the door should, as far as possible, be kept between the person of the operator and the chamber contents.
2. Wear insulated gauntlet gloves and a visor when unloading.
3. Avoid mechanical and thermal shocks (from cold, draught, etc) to the load.
4. When possible leave large loads, or loads made up of large unit volumes, overnight in a locked autoclave to cool.
5. Use containers for culture media as small as compatible with efficiency and convenience in use.

From The Public Health Laboratory Service Subcommittee on Laboratory Autoclaves Report (1978). Autoclaving practice in microbiology laboratories. *Journal of Clinical Pathology*, **31**, 418–22. Reproduced by permission of the editors of the *Journal of Clinical Pathology*.

Loading and unloading autoclaves, especially the vertical type, can induce back strain. Horizontal models are to be preferred and it is best to load and unload them with a trolley. Large modern autoclaves have integral trolley-loading facilities.

Although unpleasant odours are not a hazard they may be a consequence of autoclaving large amounts of pathological material. Such smells tend to permeate a whole building and unless there is proper ventilation they can be a source of annoyance to many people.

2.10.2 Microscopy

2.10.2.1 *Optical microscopy*   There has been concern about the frequency of reports of exploding mercury vapour lamps used as a source of ultraviolet (uv) radiation in fluorescence microscopy. These have a high mercury vapour pressure (up to

7090 kPa). An explosion presents two hazards – flying glass and the release of mercury (albeit a small amount). It has also been claimed that eyes might be damaged by the uv radiation. Some manufacturers provide uv-absorbing shields to protect the eyes. A number of incidents have been investigated, the literature searched and manufacturers have been consulted. As a result the DHSS (1985) has published safety precautions for fluorescent microscopy. Table 2.6 is based on this document which also gives some advice on the use of xenon lamps.

TABLE 2.6 Mercury vapour and xenon lamps used in fluorescent microscopy: some recommended safety measures

1. Operate fluorescence microscopes in well ventilated rooms.
2. If a mercury vapour lamp explodes increase ventilation to dilute mercury vapour.
3. Wear protective goggles or safety mask and gloves when handling new and used lamps.
4. Do not remove the plastic cover of a new xenon lamp until it has been fitted into the lamp mounting ready for insertion into the lamp housing.
5. To destroy a burned-out xenon lamp wrap it, without its plastic cover, in a large piece of thick cloth, e.g. heavy canvas. Place it on a hard surface and smash it with a hammer. Wear goggles.
6. Dirt and grease deposited during handling impair efficiency and increase risk of explosion. Allow lamp to cool, clean it with alcohol and dry with lens tissue.
7. Make electrical connections according to manufacturer's instructions. If a cooling attachment is fitted to a lamp terminal ensure that neither it nor any lamp connection is in contact with the lamp housing at any setting or the lamp may explode. The minimum clearance should be 4 mm.
8. With 200 W ultra high pressure mercury lamps ensure that the mirror image of the arc is not focussed on the electrodes or the quartz glass envelope. This may occur when the image of the discharge arc and the reflection overlap. It will cause localized overheating, shortening the life of the lamp and if an image is formed on the quartz glass this may crystallize and the lamp will explode.
9. Allow the lamp to cool for 15 minutes after switching off and before opening the lamp housing. Exposing a hot lamp to air may cause it to explode.
10. Do not run a lamp longer than the manufacturer's recommended time. Check hours of use with a lapsed time meter or log book. Frequent switching on and off shortens the life of the lamp.

Nyman (1984) studied the changes in eye function and the incidence of eye strain during two different microscopy-intensive working routines, i.e. 6.75 hours of full-time microscopy and 4 hours of microscopy during an 8-hour working day. A control group was not engaged in microscopy. Before and after the study all microscopists and controls were examined by an ophthalmologist. Visual acuity, refraction and the use of corrective spectacles were investigated. Nyman's conclusions were that intensive microscope work may give rise to subjective and objective manifestations of eye strain.

It has also been claimed that there is sometimes a smell of ozone in the proximity of uv microscopes. Proper ventilation will take care of this.

2.10.2.2 Scanning and transmission electron microscopy
Thurston (1978) reviewed the hazards of scanning electron microscopy. These include fire, toxic chemicals and explosions associated with critical drying equipment. Johansen (1984) gives more details and covers implosion hazards presented by evaporators, freeze dryers, freeze fracture and spatter-coater units.

2.10.3 Equipment with heating facilities and equipment run unattended

Twenty-seven over-heating incidents in hospital laboratory equipment, some leading to fires, were investigated by Kennedy (1985). Histology equipment was involved in 21 of them. A study allowed a strategy for fire prevention to be developed. In particular, Kennedy identified three areas where equipment is particularly vulnerable to fire and noted the aggravating factors. In one incident process equipment with a heating facility had to be left running at night and at weekends and holiday periods. Thus the failure of an electrical component was serious because no one was there to detect a 'prefire condition' such as smoke, except by chance. The histology equipment was used in premises where there were flammable solvents and wax. There was a tendency for molten wax to infiltrate into the electrical components so that the risk of fire was increased.

Kennedy found that some older equipment did not comply with modern safety standards. In such equipment, there is not likely to be a thermal safety cut-out or back-up thermostat, there is always the possibility that a fuse has been replaced by one with an incorrect rating, and there is a likelihood of component failure.

Kennedy summarized his safety strategy as follows.

Equipment should:

1. comply with a recognized standard, e.g. ESCHLE (DHSS, 1986);
2. be subject to a regular safety inspection;
3. be subject to a replacement policy (DHSS, 1982a);
4. be switched on and left unattended only when strictly necessary;
5. be backed up by appropriate fire precautions and security measures if it is left switched on unattended (DHSS, 1986).

Hinberg *et al.* (1982) calculated that in a poorly ventilated cytology laboratory where three tissue processors were in use, the amount of toluene vapour in the air exceeded 100 ppm - the maximum allowable exposure level. Evaporation of the toluene was enhanced by the wax baths maintained at 56°C placed next to the toluene baths. They considered that tissue processors should be placed under appropriate fume hoods or fume expellers unless they had integral facilities to deal with solvent vapours.

2.10.4 Hazards of fume cupboards and microbiological safety cabinets

These devices are intended to protect the user from chemical and biological hazards respectively. They are not interchangeable but have certain common features. It is ironic that some of each have proved to be intrinsically hazardous, especially when they have been constructed so that the air flow is too low to retain the agents that they are supposed to contain.

One modern fume cupboard was so constructed that the sash cable could jump from its pulley. The fail-safe device did not work because it was incorrectly fitted. Consequently the sash could easily descend suddenly and injure the arms of the user.

Some years ago the glass front panel of a microbiological safety cabinet dropped out on to the user's hands. The panel was mounted in a metal frame that extended to the top and both sides but not to the bottom edge. It was fixed with epoxy resin but the fixing was unsound. Recently there were three incidents where the front glass screen of Class I safety cabinets, all made by the same company, shattered spontaneously. There was a loud report and many fragments of glass were scattered. Fortunately most were retained within the cabinet but the users were at a considerable risk. The screen was secured by four retaining fasteners which, if not correctly adjusted, caused unnecessary stress in the glass. On two occasions it was suspected that as the 'night door' was in position and the cabinet was working under negative pressure additional stress was imposed on the screen. Users of similar cabinets were advised by the STD to have glass screens replaced by those made of Perspex.

Kennedy (1979b) reports an incident when an explosion occurred in a Class I microbiological safety cabinet that was being decontaminated. Formalin was being boiled over a micro-bunsen and the cabinet was sealed. Eventually, when the oxygen had been used up the burner went out. This was not noticed at the time and a concentration of gas built up. There was an explosion when an attempt was made to relight the burner.

Safe methods of disinfecting cabinets with boiling formalin or formalin–permanganate mixtures are reviewed by Collins (1988; see also Section 6.13.6). If naked flames for sterilizing inoculating loops are used in microbiological safety cabinets only microbunsens should be used and the gas supply should be interlocked so that it is turned off when the air flow stops. Normal size bunsens interfere with the air flow and can introduce a fire hazard. Similarly, uv sources should be interlocked.

2.10.5 Centrifuge hazards

Centrifuges are the workhorses of clinical and biomedical laboratories and it is not surprising that there have been a number of accidents and incidents involving them. The Howie Code of Practice (DHSS, 1978) states that mechanical safety is the prerequisite of microbiological safety. According to the British Standard (BSI, 1980) 'The biggest threat to the user, and to others in the vicinity, comes from (centrifuge) rotation assemblies. The most important safety requirement for a centrifuge is the provision of a strong guard barrier surrounding the rotation assembly to prevent any debris resulting from disruption being projected into the laboratory. Accidents caused by the breaking up of centrifuge heads often have explosive violence. The guard barrier of an ultracentrifuge for example, may be required to contain an energy release equivalent to the head-on collision of a car having a mass of 1 tonne (1000 kg) travelling at 160 km/h and the design has to provide for the absorption of all the kinetic energy contained in the disrupted parts.'

Uldall (1974) outlines centrifuge accidents in Danish clinical chemistry laboratories between 1954 and 1973. They include seven incidents in which parts were ejected from the centrifuge and a case of permanent injury to a finger used in an attempt to stop the rotor. The British Standard (BSI, 1980) requires that it is not possible for the motor to be energized unless the lid is closed (except where the manufacturer supplies zonal or continuous flow centrifuge head machines, for which there are special safety requirements).

Some DHSS centrifuge accident statistics are shown in Table 2.7. In over 15 years, despite the large numbers of incidents reported and apart from electrical shocks (see Table 2.1) only two minor injuries were reported. Neither of these was caused by the ejection of moving parts. Such incidents can have serious consequences however, and in one accident in an industrial laboratory the user lost an eye because the centrifuge casing failed to contain a part that had become detached from the rotor.

While some of the NHS incidents were caused by faulty components, others were the result of misuse. Kennedy (1979a) observed that the most common centrifuge accident was the 'spin-off' type where buckets and trunnions hit the casing after being spun off the rotor. He considers that the commonest causes of centrifuge accidents in the NHS are:

1. Failure to balance the load.
2. Failure to locate the trunnions and buckets properly, sometimes because the centrifuge is placed too high for the user to see inside the bowl.
3. Too rapid an acceleration rate, i.e. over-enthusiastic use of the speed control.

The tendency of some obsolescent centrifuge casings to fail when struck by objects from inside led the DHSS to advise NHS users to scrap them in the interests of safety.

Rotor failure on any type of centrifuge is usually initiated by cracks or fissures. These may be caused by corrosion or mechanical damage. Corrosion usually starts at high stress areas and often in places where dirt accumulates; the bottoms and sides of buckets and cups are common sites. Pitting occurs when the anodizing of rotors is lost and it is important not to use alkaline solutions which attack anodized areas. Corrosion was a feature of some NHS centrifuge incidents. In some cases saline used for red cell washing was responsible, and in another oxalic acid,

TABLE 2.7 Some DHSS centrifuge incident statistics

Type of incident	Number
Period of study: February 1971 – October 1986 121 incidents were investigated, including:	
Ejection of buckets and other parts	17
Ejection of glass fragments from broken tubes	2
Centrifuge moved bodily as a result of impact inside	3
User cut wrist on sharp metal	1
User injured by parts of spring-loaded lid lock assembly	1

As a result of these investigations:

1. Four Hazard Notices and three Safety Information Bulletins were issued.
2. Two Health Equipment Information papers were published advising on the need for instruments to be modified in the interests of safety and recommending that obsolete instruments with a poor safety record should be scrapped.

sodium hydroxide and sulphuric acid that had been used for preparing material for culturing tubercle bacilli.

The safety and care of ultracentrifuge rotors are reviewed by Dewhurst (1976). He points out that although the potentially lethal nature of rotor failure is fully appreciated and is the basis for a certain amount of mistrust among laboratory workers, there are few grounds to support such an attitude. He states that in the ultracentrifuge the need for high G forces means that the rotor is loaded high in the elastic region and stretching occurs. A rotor will withstand only a finite number of stress cycles before it fails through metal fatigue. In addition, any mechanical changes or severe corrosion in the rotor can cause localized areas of high stress which lead to premature failure. Rotors may be derated however, and used at lower speeds when they are approaching the end of the safe working life at high speeds. When this is done the safety device that prevents the rotor from running at too high a speed should be changed to prevent accidental over-run. Suitable speed discs to cover derating are available from manufacturers, who will also inspect rotors.

Dewhurst (1976) advocates the following precautions for the care of ultracentrifuge rotors:

1. Proper cleaning procedures.
2. Use of a brush that does not damage anodizing.
3. Correct storage, i.e. disassembled and upside-down.
4. Adherence to manufacturers' instructions about overloading and speed.

Accidents in centrifuges used for infectious materials may result in the massive release of aerosols containing infectious particles (Collins, 1988 and Section 6.8.1.7). For example, Hunter (1971) recalls that in 1905 a suspension of glanders bacilli was being centrifuged when a failure occurred and the organisms were scattered. Four people were infected and three died. Helwig (1940) reports a case in which a highly concentrated chick embryo culture of Western equine encephalitis virus was being centrifuged. The machine was new and had not been used before. Apparently too much power was applied (see above!) and the virus suspensions were ejected and sprayed all over the walls and windows. The operator was infected and died. Wedum (1973) cites 11 cases where infection or hypersensitivity resulted from centrifuge accidents.

Chemical explosions have also been associated with centrifuge failures. The Medical Research Council (MRC, 1984) cites examples investigated by the Beckman company, which is probably the world's largest manufacturer of centrifuges. These explosions followed the break-up of rotors that were running at excessive speeds, possibly after failure of the electronic speed control systems. The rotors disintegrated into finely divided metal particles. Thus a large area of unoxidized titanium metal was exposed to water derived from the load and Freon gas when the cooling coils were disrupted by flying debris. Sparks from the disintegration of the motor triggered off an explosion which blew the chamber door off the instrument and severely damaged the laboratory. As a result of their investigations Beckman derated some rotors and 'retrofitted' new electronics. They also advised that certain rotors, over ten years old, should not be used.

2.11 SOME GENERAL OBSERVATIONS

2.11.1 Design and construction of equipment

Kennedy (1979a) stated that designers should bear in mind that in general their instruments will be used by ordinary people who will be more interested in getting results than in technological innovations, quality of engineering and the like. In some circumstances these people will be working under conditions resembling those of a factory line and may, justifiably, be in a hurry. Nothing can be made foolproof of course, but pitfalls in operation can be minimized and user maintenance can be simplified. In a review of the usefulness of industrial accident data Kletz (1976) observed that it is not only equipment that fails: men also fail. He states 'If the failure rate is unacceptable we must re-design the equipment – it is no use telling a man to be more careful. We might just as well reprimand a light bulb for going out.'

A sound approach to design, especially where there is likely to be misuse, or a hazard may be anticipated, is to build in an interlocking device which makes the user physically incapable of exposing himself to danger unless he deliberately by-passes the mechanism. Interlocking is a requirement of some British Standards. Where recognized standards exist, these should be followed. For electrically operated equipment the all-embracing standard is ESCHLE (DHSS, 1986). In addition to electrical safety this covers protection against mechanical hazards, mechanical resistance to stress, impact and vibration, equipment temperature limits and protection against fire.

2.11.2 The *UK Health and Safety at Work etc. Act 1974*

Section 6 of this Act imposes a general duty upon any person who designs, manufactures, imports or supplies articles for use at work to ensure, so far as is reasonably practicable, that the article is so designed and constructed as to be safe and without risks to health when properly used. He is also required to carry out, or arrange for the carrying out, of such testing and examination as may be necessary for the performance of the duty imposed upon him by the Act. Furthermore, he must provide adequate information in connection with the use of the article at the place of work for which it has been designed and tested and about any conditions necessary to ensure that when in use it will be safe and without risks to health. Similar duties are imposed upon any person who erects or installs any article for use at work in any premises where that article is to be used by persons at work. There is general agreement that this legislation is difficult to enforce in practice and plans are being made to revise it.

2.11.3 Management of equipment

The DHSS document on the Management of Equipment (DHSS, 1982a) states in its Introduction that 'During the past few years, investigation of serious accidents has continued to reveal shortcomings in the management of equipment use in hospitals and associated establishments. Common factors in these accidents include the following:

1. Inferior quality or worn-out equipment.
2. Inadequacies or mistakes in servicing.
3. Use of unsuitable or incompatible ancillary equipment.
4. Inadequate knowledge of or training in the use of the apparatus or system involved.

A particularly disturbing feature has been the recurrence of accidents with similar causes in spite of earlier warnings that describe the accidents in considerable detail and suggest how they might be avoided. The aim of the publication is to recommend a system of equipment management which, when fully implemented, will ensure that all equipment used in the NHS is not only suitable for its purpose, but is also maintained in a safe and reliable condition, is understood by the user and can be used with confidence.'

There are five activities which are essential features of equipment management:

1. Selection of equipment.
2. Acceptance procedure.
3. Training.
4. Servicing (maintenance, repair and modification).
5. Replacement policy.

Each of these is discussed in detail in the Management of Equipment document (DHSS, 1982a).

The various mishaps in clinical and biomedical laboratories that have been described above will amply justify these features. It is therefore recommended that the principles of equipment management are adopted in all such laboratories.

2.12 ACKNOWLEDGEMENTS

I wish to thank my colleagues in the STD for their assistance; R.W.B. Allen, A. Horn, N. Jennings and R.F. Latchford for their contributions to the investigations that provided material for this chapter; G.R. Higson (Technical Director, STD), D. Hurrell, C. Robertson and D. Wells for their suggestions in improving the chapter. I also wish to thank Dr M.A. Buttolph, Safety Advisor to the University of London, for drawing my attention to some of the references that have been cited. All opinions expressed in this chapter are those of the author and do not necessarily reflect those of the DHSS.

2.13 REFERENCES

APEX (1985) *New Technology: A Health and Safety Report.* London: Association of Professional, Executive, Clerical and Computer Staff.

British Cryogenics Council (1970) *Cryogenics Safety Manual: A guide to good practice. Part IV, Hydrogen separation plants.* London: British Cryogenics Council. p. 76.

BSI (1955) *BS 2646. Copper laboratory autoclaves.* London: British Standards Institution.

BSI (1966) *BS 3970. Part 1: Sterilizers for porous load. Part 2: Sterilizers for bottled fluids.* London: British Standards Institution.

BSI (1973a) *BS 349. Identification of contents of industrial gas cylinders.* London: British Standards Institution.

BSI (1973b) *BS 349C. Chart for the identification of contents of industrial gas cylinders.* London: British Standards Institution.

BSI (1980) *BS 4402. Specification for safety requirements for laboratory centrifuges.* London: British Standards Institution.

BSI (1983) *BS 4803. Guide on protection of personnel against hazards from laser radiation.* London: British Standards Institution.

Collins, C.H. (1988) *Laboratory-Acquired Infections.* 2nd edn. London: Butterworths.

Dewhurst, F. (1976) Safety and care of ultracentrifuge rotors. *Laboratory News,* March 30.

DHSS (1978) *Code of Practice for the Prevention of Infection in Clinical Laboratories and Post-mortem Rooms.* London: HMSO.

DHSS (1979) The use and care of anaerobic culture equipment. *Health Equipment Information No 76.* London: Department of Health and Social Security.

DHSS (1980) Sterilizers. *Health Technical Memorandum 10.* London: HMSO.

DHSS (1982) Danger of explosion in freeze-drying equipment and vacuum systems used to process materials containing sodium azide. *HN (Hazard) (82)10.* London: Department of Health and Social Security.

DHSS (1982a) *Management of Equipment. Health Equipment Information No. 98.* London: Department of Health and Social Security.

DHSS (1982b) Fire Safety in Health Care Premises. *Health Technical Memorandum No 83.* London: HMSO.

DHSS (1984a) Review of four wax-embedding centres that comply with ESCHLE. *Health Equipment Information No. 123.* 36/84 London: HMSO.

DHSS (1984b) Storage of medical, pathology and industrial gas cylinders. OPS.1WG. London: Department of Health and Social Security.

DHSS (1984c) Guidance on the safe use of lasers in medical practice. London: Department of Health and Social Security.

DHSS (1985) Prevention of explosions and some safety measures in the use of mercury and xenon lamps that are used in fluorescence microscopy. SIB(85) 55. London: Department of Health and Social Security.

DHSS (1986) Electrical Safety Code for Hospital Laboratory Equipment, ESCHLE. *Health Equipment Information No. 158.* London: Department of Health and Social Security.

DOE (1980) Refrigerated storage cabinets for volatile flammable liquids. *Standard Specification (M&E) No. 157.* (UK only). London: Department of the Environment

Ederer, R.M., Tucker, B. and Vikmanis. A. (1971) Accident facts in two clinical laboratories: A ten year study. *Journal of Chemical Education* **48**, A91–A93.

Edmunds, O.P. and Sutton, M.L. (1985) Radiation burn in a university laboratory. *Journal of Social and Occupational Medicine* **35**, 118–120.

Everett, K. and Graf, F.A. (1971) Handling perchloric acid and perchlorates. In *Handbook of Laboratory Safety* (ed. N.V. Steere). Boca Raton FA: CRC Press. pp. 265–276.

Gill, F.S. (1980) Heat. in *Occupational Hygiene* (eds H.A. Waldron and J.M. Harrington). London: Blackwell. pp.225–256.

Haddon, W., Suchman, E.A. and Klein, D.A. (1964) *Accident Research: Methods and Approaches.* New York: Harper and Row.

Hamilton, M. (1984) Glass hazards in laboratories. *Safety Management* **10**, 31–37.

Hellings, R.O. (1981) Health and safety issues in universities and other institutions. In *Occupational Health and Safety Management* (eds S.S. Chissick and R. Derricot). London: John Wiley.

Helwig, F.C. (1940) Western equine encephalitis following accidental inoculation with chick embryo virus. *Journal of the American Medical Association* **115**, 291–292.

Hinberg, I., Katz, L., Weber, F. and Sullivan, D. (1982) Hazard warning: automatic tissue processors. *Dimensions*, April, p.17.

HSE (1976) *Plant inadequately guarded.* Health and Safety: Manufacturing and Service Industries, New Entrants. 5, p.13. London: Health and Safety Executive.

HSE (1984) Health effects of VDUs (a bibliography compiled by Rosemarie Thomas). London: Health and Safety Executive.

HSE (1985) *Approved Code of Practice for the protection of persons against ionising radiations arising from any work activity.* London: Health and Safety Executive.

HSE (1986) *Working with VDUs.* London: Health and Safety Executive.

Hunter, D. (1971) Accidents in bacteriological laboratories. In his: *The Diseases of Occupations.* London: English Universities Press. p.1160.

IEC (1984) *Effects of current passing through the human body. Part 1: General Aspects.* Publication No. 479–1. Geneva: International Electrotechnical Commission.

ILO (1986) *Conditions of Work Digest - Special issue on VDUs. Vol 5 (1).* Geneva: International Labour Organisation.

Johansen, B.V. (1984) Hazards related to EM-laboratory instrumentation. *Ultrastructural Pathology* 219–225.

Kennedy, D.A. (1979a) Investigation of some accidents involving laboratory instrumentation. *Laboratory Practice* **28**, 17–19.

Kennedy, D.A. (1979b) Disinfection of microbiological safety cabinets (Letter), *Gazette, Institute of Medical Laboratory Sciences* **24**, 346.

Kennedy, D.A. (1985) Electrical equipment fires: causal factors and a strategy for prevention. *Laboratory Practice* **34**, 90–93.

Kibblewhite, F.J. (1984) Analysis of university accidents. *The Safety Practitioner* **2**, 13–21.

Kletz, T.A. (1976) Accident data - the need for a new look at the sort of data that are collected and analysed. *Journal of Occupational Accidents* **1**, 95–105.

Maley, M.P. (1986) Burns from microwave oven. *Lancet* i, 1147.

Matteson, M.T. and Ivancevitch, J.M. (1982) Stress and the medical technologists. II. Sources and coping mechanisms. *American Journal of Medical Technology* **48**, 169–176.

Minok, G.Y., Waggoner, J.G., Hoofnagle, J.H. *et al.* (1982) Pipetter's thumb. *New England Journal of Medicine* **306**, 751.

MRC (1984) Ultracentrifuge explosion hazard. Health Hazard Note No. 74. London: Medical Research Council.

MRC (1985) Explosion hazard: use of microwave ovens for media preparation. Health Hazard Note No 81. London: Medical Research Council.

Nyman, K.G. (1984) An experimental study on visual strain and microscopy work. *Acta Ophthalmologica.* (Supp) **161**, 94.

Pearce, B.G. and Shackel, B. (1979) The ergonomics of scientific instrument design. *Journal of Physics, E. Scientific Instruments* **12**, 447–54.

PHLS (1978) Autoclave practice in microbiology laboratories. Report of a survey. *Journal of Clinical Pathology* **31**, 418–422.

Steere, N.V. (1980) Physical, chemical and fire safety. In *Laboratory Safety: Theory and Practice* (eds. A.A. Fuscaldo, B.J. Erlick and B. Hindman). New York: Academic Press. pp. 4–27.

Thurston, E.L. (1978) Health and safety hazards in the SEM laboratory: Update 1978. *Scanning Electron Microscopy* **2**, 849–853.

TUC (1985) *Guidelines on VDUs*. London: Trades Union Congress.

Uldall, A. (1974) Occupational risks in Danish clinical chemistry laboratories. *Scandinavian Journal of Clinical Laboratory Investigation* **33**, 21–25.

Wedum, A.G. (1973) Microbiological centrifuging hazards. In *Centrifuge Biohazards: Proceedings of a Cancer Research Safety Symposium*. Frederick, Maryland: Cancer Research Center. pp. 5–16

Weeks, J.L. (1976) Radiation exposure in the laboratory. *Journal of Social and Occupational Medicine* **26**, 9–12.

Willhoft, T. (1986) Victims of the pop bottle. *New Scientist*, 21 August, pp. 28–30.

Williams, A.R. (1985) Uses of ultrasound and their hazards. In *Handbook of Laboratory Health and Safety Measures*. (ed. S.B. Pal). Lancaster: MTP Press. pp. 299–326.

3 CHEMICAL HAZARDS

J.F. Stevens

Many different chemicals are used in clinical and biomedical laboratories and because many of them are potentially hazardous it is essential that laboratory workers know enough about their properties to handle them safely. Health hazards may result from contact with skin or eyes, by accidental ingestion, e.g. in mouth pipetting, or by inhalation of gases, vapours or fine powders.

The new *Control of Substances Hazardous to Health Regulations (COSHH) 1987* places a duty on employers and employees (among other things) to handle such substances with due care and to assess the risks of exposure to chemicals and other agents.

In a work of this nature it is not possible to mention every hazardous or potentially hazardous chemical that a clinical laboratory worker might encounter and therefore only those that are relatively common or topical are included. Fortunately reference works are available and laboratory managers and safety staff would be well advised to have them close at hand. Particularly useful sources of information are the books of Bretherick (1979, 1981) and Croner's *Substances Hazardous to Health* (1986). Much useful information is also given by Chivers (1985) and Smith (1985). The registry of toxic effects of chemicals (NIOSH, 1983) and the guidance note of the Health and Safety Executive on Occupational Exposure Limits (HSE, 1984) are also useful. In addition, some manufacturers' catalogues are also valuable sources of information. The Department of Health and

Social Security (DHSS), through its Supplies Technology Division (STD), issues Hazard Warning notices about chemical hazards which are generated in response to particular accidents or incidents and are designed to prevent recurrence elsewhere. These warning notices should be circulated among senior laboratory staff, not filed away. Some more specialized references are mentioned below.

3.1 HAZARDOUS CHEMICALS: SOME DEFINITIONS

Poisonous, toxic, corrosive and irritant chemicals now come within the orbit of COSHH but at present there are also regulations about certain other substances. These include scheduled poisons, carcinogens and radioactive substances.

3.1.1 Scheduled poisons

Some of the chemicals used in laboratories are subject to the *Poisons Act 1972*, the *Poisons Rules 1978* and the *Poisons List Order (Schedule I or II, 1978)*. These substances are usually referrred to as 'Scheduled Poisons' and most of them *must* be kept in designated and locked cupboards. Full lists and regulations are kept in all pharmacies.

3.1.2 Carcinogens, teratogens and mutagens

The use of these is covered by the *Carcinogenic Substances Regulations*, (1967, amended 1973). Useful definitions of such substances are:

- *Carcinogen* A substance that competent authorities agree can cause cancer in man or animals at a specified level of exposure. The 'incubation period' after exposure may be as long as 40 years (MRC, 1981).
- *Suspect carcinogen* A substance not known to be a carcinogen but which has a close structural similarity to one. There may

be some evidence that the substance is carcinogenic for animals (MRC, 1981).

• *Teratogen* A substance that causes foetal abnormalities at concentrations less than those that cause embryonic death (Cater and Hartree, 1977)
• *Mutagen* A substance that causes irreversible chemical aberrations in chromosomal nucleic acid. (Cater and Hartree, 1977).

Advice is given by the Chester Beattie Research Institute (1977) and the Medical Research Council (MRC, 1981).

3.1.3 Radioactive substances

The various radionuclides used in clinical and biomedical laboratories are subject to the *Radioactive Substances Acts 1948 and 1960*. New regulations are expected. Various codes of practice have been published e.g. by the Health and Safety Executive (HSE, 1981). The main hazards are contamination of the skin, deposition in the body, beta and gamma radiation and spillage of radioactive wastes. These are compounded by improper or inadequate disposal of waste.

3.2 PERSONAL PRECAUTIONS WHEN HANDLING CHEMICALS

These have been considered in general in Chapter 1 but are emphasized here because serious injuries may result if chemicals are splashed on the skin and into the eyes, spilled on any part of the person, accidentally ingested or their vapours inhaled.

3.2.1 Laboratory clothing

The Dowsett–Heggie type (see Section 1.7.1) is the best for normal work and should be supplemented with plastic aprons for work with corrosive chemicals, even when merely filling stock bottles from larger storage containers.

3.2.2 Safety spectacles or vizors

These should always be worn for handling strong acids and alkalis, ammonia, powerful oxidizing agents and disinfectants. They should also be worn when chemicals are ground in mortars and during the spraying of chromatograms.

3.2.3 Gloves

Heavy duty rubber gloves should be worn when hazardous chemicals are poured from stock into smaller containers and for dealing with spillages. Disposable gloves should be worn for laboratory work with such chemicals and for handling radioactive and carcinogenic substances.

3.3 STORAGE OF CHEMICALS

Many of the hazards associated with chemicals may be obviated by correct storage.

3.3.1 The chemical stores

In the interests of safety and fire prevention chemical stores should be away from the main building. They should not be too far away however, or there will be a tendency for the staff to keep larger amounts of chemicals in the laboratory than are reasonable or safe.

The building should be cool and fireproof. If potentially explosive materials are stored, the walls should be thick and the roof comparatively thin to avoid a horizontal blast if there is an accident. The floor should be of concrete and leakproof and there should be a sill at the threshold so that spilled chemicals are retained in the store and cannot leak out under the door. Shelves may be of wood or metal but slate is preferable.

An alphabetical arrangement of chemicals may be most unsafe. It may result in acids and organic solvents being stored close together - a fire hazard, illustrated by a recent incident when fuming nitric acid was stored next to acetone. There are strong arguments in favour of separate 'flammable stores'. Some incompatible chemicals, which should not be placed close to one another, are listed in Table 3.1. Advice on the storage of flammable chemicals is given by HSE (1977).

Periodic checks should be made of the state of chemicals, e.g. for leakage, corrosion of containers and for dates of purchase and expiry dates. The latter are particularly important with some chemicals. They indicate that the substance may become potentially hazardous if it starts to decompose.

TABLE 3.1 Some incompatible chemicals

Acetic acid	Chromic acid, nitric acid, OH-compounds, ethylene glycol, perchloric acid, peroxides and permanganates
Acetone	Concentrated sulphuric and nitric acids
Bromine	Ammonia, butadeine, butane and finely divided metals
Chromic acid	Acetic acid, naphthalene, alcohol, glycerol and flammable liquids
Copper	Hydrogen peroxide
Cyanides	Acids and alkalis
Flammable liquids	Chromic acid, hydrogen peroxide, nitric acid and halogens
Iodine	Ammonia
Oxalic acid	Silver and mercury
Picric acid	Copper and lead
Potassium permanganate	Glycerol and sulphuric acid
Sodium azide	Lead and copper (plumbing!)
Sulphuric acid	Chlorates, perchlorates, permanganates and water

3.3.2 Transport of chemicals

Bottles of chemicals, especially winchesters, should be transported in bottle carriers which are available commercially in various sizes. Bottles, again especially winchesters, should not be carried or even lifted by the neck. One hand should always be placed under the base. There have been many accidents in which the neck broke away from the bottle, spilling the contents. In one, a man's legs were severely burned by concentrated sulphuric acid.

3.4 GENERAL PRECAUTIONS IN THE LABORATORY

Shelves and storage cupboards used for chemicals in the laboratory should not be in direct sunlight, nor should they be over or near to radiators or sources of heat. Hazardous chemicals should not be placed on high shelves. Recommendations for storing certain chemicals are given in Section 3.5. All bottles and containers should be boldly labelled and display the appropriate warning signs (see colour plate section). Ideally the number and quantities of chemicals kept in the laboratory itself should be minimal. The proper place for bulk chemicals is the chemical store. It is inadvisable to keep more than 500 ml of strong acids, alkalis and flammable solvents in the laboratory although it is appreciated that problems arise when chemical stores are far from the laboratory itself. Chemicals should not be left on benches but returned to shelves or cupboards after use. Care should be also taken that 'incompatibles' (Table 3.1) are not placed close together.

Volatile and flammable liquids, e.g. organic solvents, should be kept in the steel flammable liquid cabinets that are now generally available. They should not be stored in ordinary refrigerators as these do not have spark-proof electrical controls (see Section 2.9.3 and Table 2.1).

The benches should be kept tidy, with the minimum of clutter. Crowded benches foster accidental spillages.

There should be no mouth pipetting, even of water. Errors do occur and a general ban will avoid the possible ingestion of a harmful liquid that 'looked like water'.

Flammables should be handled well away from naked flames.

Flammable compressed gas cylinders (including LPG) should be well away from naked flames and any other source of heat.

Bottles and containers should be opened with caution, especially those with Hazchem labels. The contents may be under pressure.

'Empty' containers should be handled with caution. Residues may be hazardous. They should be rinsed out (but see Disposal, Section 3.7) before return to suppliers.

3.5 HAZARDS AND PRECAUTIONS WITH SOME CHEMICALS USED IN CLINICAL AND BIOMEDICAL LABORATORIES

As indicated above, this list cannot be complete. For substances not listed see Bretherick (1979, 1981), HSE (1984), NIOSH (1985) or Cronor (1986).

- Acids are toxic and corrosive. Store at low level in a cool place, preferably on plastic drip trays. Dilute with care, and always add acid to water. Wear goggles or vizor. See under hydrochloric, nitric and sulphuric acids below.
- Alkalis are corrosive. Store at low level in cool. Weigh solids with care.
- Ammonia is highly toxic especially to eyes and lungs. Open bottles in well ventilated surroundings or in a fume cupboard. Wear goggles or a vizor and a plastic apron.
- Aromatic amines are carcinogens. *Do not use.* See benzidine, *o*-dianisidine, *o*-tolidine.
- Arsenic is a Scheduled Poison.
- Auramine is a Controlled Substance*. Keep exposure to minimum.
- Azides form explosive compounds with metals. *Do not use.* (See Section 2.3.1).
- Benzene is hepatotoxic and carcinogenic. Do not inhale and use only in small volumes in fume cupboard.
- Benzidine is a carcinogen. Use is prohibited.

- Bromine is highly toxic. Wear a vizor and use fume cupboard.
- Carbon dioxide (solid) can cause serious burns to skin. Store unsealed (pressure builds up in closed containers) in a ventilated room, otherwise there is danger of asphyxiation. Do not carry in closed lifts. Cover with a cloth when breaking up blocks.
- Carbon tetrachloride is toxic and suspected carcinogen. *Do not use.*
- Cyanides are Scheduled Poisons. They have a rapid effect. Antidote kits are not usually effective unless used immediately.
- Cyanogen bromide is highly toxic. Store in a cool place and surround bottle with an outer container packed with kieselguhr. Use in fume cupboard and neutralize after use with ammonia.
- Cytochalasin is mutagenic. Use with great care.
- *o*-Dianisidine is a Controlled Substance*. Use should be restricted.
- Dichlorbenzidine is a Controlled Substance*. Use should be restricted.
- Diethyl ether is highly flammable with a low flash point. It is used as an anaesthetic and is known to be addictive. Unless steps are taken by the manufacturer to prevent oxidation, peroxides may build up and there is then an explosive risk if evaporated to dryness or distilled. Ozonic ether, at one time used in blood detection, is particularly hazardous (Kennedy & Jennings, 1985). Store all ethers in a cold place in a dark bottle but not in an unmodified refrigerator (see Section 2.9.3.1). Use as solvent only if there is no alternative.
- Formaldehyde is toxic to eyes and lungs. Wear goggles and do not inhale. There is no evidence to suggest that concentrations used in laboratories are a carcinogenic hazard (Clark 1983b). Keep in a well ventilated room.
- Hydrochloric acid is toxic and an irritant. Do not inhale vapour. Use goggles or a vizor and wear gloves. To dilute add acid to water.

- Hydrogen peroxide is toxic. Concentrations of more than 50 vol% may be explosive. Store at 4°C on the lowest shelf of a refrigerator and loosen the cap to release pressure periodically.
- Hypochlorites – see sodium hypochlorite
- Mercury releases a toxic vapour. Keep in sealed containers in a cool and dark place. Some salts are Scheduled Poisons. Clean up escapes from equipment promptly – even small amounts.
- Naphthylamine. Beta-naphthylamine is carcinogenic. Alpha-naphthylamine is a Controlled Substance. *Do not use either.*
- Nitric acid is toxic and an irritant. Store in a cool place. Open fuming nitric acid bottles periodically to release pressure.
- Nitrogen (liquid) may burn the skin, particularly if clothing is contaminated. Avoid touching metal objects exposed to the liquid. There is a danger of asphyxiation unless it is kept in a well ventilated room. Store in an open container to avoid pressure build up. Use only Dewar flasks (see Section 2.9.3.2). Use a vizor and gloves when pouring.
- Osmium tetroxide is very toxic and an irritant. Use in a fume cupboard.
- Perchloric acid is highly toxic and corrosive. It is potentially explosive. Keep only small amounts and store away from solvents and acetic anhydride. If wood or brickwork becomes contaminated get assistance from the National Chemical Emergency Centre. Contaminated material may ignite spontaneously (see DHSS, 1976).
- Phenol as a pure substance is highly toxic and corrosive. It can be absorbed through the skin. Wear goggles and gloves. Clear phenolics used as disinfectants are less toxic but similar precautions should be taken.
- Phosphorus pentoxide is a powerful drying agent which burns skin and reacts violently with water. Wear goggles and gloves. Do not inhale.
- Picric acid is toxic and an irritant. Store under water. It forms explosive compounds with metals and if spilled on cement or concrete flooring unstable calcium picrate may be formed.
- Selenium compounds are teratogenic. Avoid inhaling the dry powder.

- Sodium hypochlorite is used as a disinfectant. Concentrated solutions are toxic and corrosive. Store in a cool and ventilated place. Bottles should have closures with integral pressure release valves. Bottles may burst due to pressure build up.
- Sulphuric acid has a powerful dehydrating action on the skin. It reacts violently with water. Add acid to water slowly and surround mixing vessel with ice water.
- Thallium compounds are Scheduled Poisons.
- *o*-Tolidine is a Controlled Substance*. *Do not use.*

3.6 SPILLAGE

Notwithstanding all precautions spillages of chemicals, particularly liquids, will happen at some time. They should be dealt with as soon as possible, but precautions must be taken and it may be necessary to delay cleaning up until the proper equipment and materials are brought to the site. The safety officer or a senior member of the staff should be informed immediately as he will usually take charge of the cleaning up procedures.

3.6.1 Emergency equipment

Ideally emergency equipment should be available in all laboratories where chemicals are handled and certainly in chemical laboratories. It should include the following, although, of course, not every item will be used for every spillage.

3.6.1.1 *Protective clothing*   This should include plastic or rubber aprons, heavy duty rubber and light plastic disposable gloves, rubber (wellington) boots. In some laboratories respirators may be necessary. These should be used only by staff specifically trained for that purpose.

3.6.1.2 *Equipment*   There should be a supply of plastic trays and buckets, plastic dust pans or disposable scoops, large forceps to pick up broken glass, mops, cloths and heavy duty paper towels.

* Controlled in the UK by the Carcinogenic Regulations 1967.

3.6.1.3 *Neutralizing agents* These will depend on the kind of chemicals used in the laboratory but should include sodium bicarbonate for neutralizing strong acids, and dry sand. Commercial spillage kits are available and these usually contain neutralizing agents for the common chemicals (acids, alkalis, solvents, mercury, cyanides etc.).

3.6.1.4 *Information* Spillage charts are available from some chemical manufacturers (e.g. British Drug Houses) but every chemical laboratory should have immediately available a copy of the *Dangerous Chemical Emergency Spillage Guide* (Warren and Potts, 1985).

3.6.2 Some common spillages

3.6.2.1 *Strong acids* Windows should be opened and appropriate protective clothing donned. Water should not be used. The spillage should be sprinkled generously with sodium bicarbonate or soda ash until all effervescence has ceased. The mess may then be swept into plastic trays or buckets, using folded paper towels, and washed down a sink or sluice with large amounts of cold water. It may be easier to clean up large spillages if sand is heaped on the area after neutralization. In this case it should be disposed of outside the building, not down a laboratory sluice.

3.6.2.2 *Alkalis, ammonia and formalin* The windows should be opened and the spillage washed liberally with cold water. If large amounts of ammonia or formalin are spilled respiratory protection will be necessary.

3.6.2.3 *Cyanides* Breathing apparatus and gloves should be worn. Material should be transported in a bucket to a safe place, neutralized with sodium hydroxide or ammonia and allowed to decompose before decanting to waste using large quantities of water. The spillage site should be cleaned with water and detergent.

3.6.2.4 *Phenol* Gloves should be worn and spillages treated with large volumes of water before being mopped up.

3.6.2.5 *Organic solvents* All naked flames should be extinguished, not only in the room where the spillage occurs, but in all adjacent rooms as vapours can be carried quite a long way, especially in artificially ventilated premises. The windows should be opened. A nonflammable dispersing (detergent) agent should be applied to the spillage to make an emulsion which can then be mopped up and washed down a sink or drain. (In some places disposal is unlawful, see Section 3.7.) Again, sand is a useful absorbent. Sawdust should not be used as it may create a fire hazard.

3.6.2.6 *Carcinogens etc.*
See Section 3.8.

3.6.2.7 *Radioactive substances*
See Section 3.9.

3.7 DISPOSAL OF CHEMICAL WASTE

There are specific legal requirements about the disposal of waste chemicals when they are discharged into the public sewers. All trade effluents are covered by the *Public Health Act, 1961* which requires consent by the Water Authority before they are discharged into sewers. The Authority is interested in the concentration of chemicals only at the point where they enter the sewer and this may be very different from that in the laboratory sink or waste system. Other legislation, the *Deposit of Poisons Act, 1972*, the *Control of Pollution Act, 1974* and the *Control of Pollution (Special Waste) Regulations, 1980* may also affect the disposal of laboratory waste (especially radioactive materials).

It is prudent and courteous to discuss the disposal of laboratory waste with the local Environmental Health Officers and the Health and Safety Executive Inspectors. Sound advice on the disposal of many chemicals is provided by MCA (1972), Bretherick (1979, 1981), Chivers (1983) and Cronor (1986).

The following recommendations seem to be reasonable:

While some chemicals may be flushed down the laboratory sink they should not be poured away one after another. For example, acids should not be allowed to mix in the laboratory drains with hypochlorites or cyanides. Large volumes of water should be flushed down between each chemical.

3.7.1 Acids, alkalis and oxidizing agents

These should be neutralized and washed down the laboratory sink with large volumes of cold water.

3.7.2 Cyanides

After neutralizing cyanides may also be washed down the sink, but large volumes of water should be flushed down the sink both before and after pouring away cyanide residues.

3.7.3 Organic solvents

If there is less than 500 ml of a water miscible solvent then it may be flushed down the sink. If there is more than 500 ml or the solvent is not miscible with water it should be placed in a properly labelled metal screw-capped container to await collection by a waste disposal company. Some solvents are incompatible and mixtures may even be explosive. Advice should be obtained (e.g. from Bretherick, 1979 and 1981). Waste should be collected regularly and frequently to avoid the accumulation of large amounts and the consequent fire risk.

3.7.4 Carcinogens
See Section 3.8.

3.7.5 Radioactive substances
See Section 3.9.

TABLE 3.2 Some carcinogenic substances that may be encountered in clinical and biomedical laboratories

aflatoxicol	N-nitrosodibutylamine
aflatoxins	N-nitrosodiethanolamine
2-aminoanthracene	N-nitrosodiethylamine
2-aminobiphenyl	N-nitrosodiisopropanolamine
1-aminonaphthalene	N-nitrosodimethylamine
2-aminonaphthalene	N-nitrosodiphenylamine
2-anthramine	N-nitrosodi-N-propylamine
arginylarginine ß-naphthylamide	N-nitroso-N-ethylbutylamine
1,2-benzanthracene	N-nitroso-N-ethylurea
benzidine	N-nitroso-N-methylaniline
benzo[a]pyrene	N-nitroso-N-methylbutylamine
benzo[e]pyrene	N-nitroso-N-methylethylamine
6-benzoyl-2-naphthol	N-nitroso-N-methylglycine ethyl ester
3,4-benzpyrene	
chloromethyl methyl ether	N-nitrosomethylurea
3,3'-diaminobenzidine	N-nitrosomorpholine
o-dianisidine	1-nitroso-2-naphthol
dibenz[a,j]acridine	2-nitroso-1-naphthol
1,2:3,4-dibenzanthracene	1-nitroso-2-naphthol-3,6-disulfonic acid
1,2:5,6-dibenzanthracene	
dibenzo[a,i]pyrene	N-nitroso-N'-nitro-N-ethyl-guanidine
3,3'-dichlorobenzidine	
diisopropanolnitrosamine	N-nitroso-N'-nitro-N-methyl-guanidine
p-dimethylaminoazobenzene	
7,10-dimethylbenz[c]acridine	N-nitroso-N'-nitro-N-propyl-guanidine
7,12-dimethylbenz[a]anthracene	
9,10-dimethyl-1,2-benzanthracene	4-nitrosopiperazine-1-carboxylic acid ethyl ester
1,4-dimethyl-2,3-benzphenanthrene	
ethionine	1-nitrosopiperidine
2-ethylanthracene	N-nitrosopyrrolidine
20-methylcholanthrene	nitroso R salt
N-(N'-methyl-N'-nitroso-[aminomethyl])benzamide	5-nitroso-2,4,6-triaminopyrimidine
	phenylalanine ß-naphthylamide
N-methyl-N'-nitro-N-nitroso-guanidine	4α-phorbol
	propidium iodide
O-methylsterigmatocystin	ß-propiolactone
morgan's base	N-propyl-N'-nitro-N-nitrosoguanidine
α-naphthylamine	
ß-naphthylamine	sterigmatocystin
4-nitrobiphenyl	o-tolidine
nitrosobenzene	
N-nitroso-N-butyl-N-propylamine	

3.7.6 Unused chemicals and old stocks

These may be hazardous particularly if they have been given an expiry date and this is long past. It is best to ask the manufacturers for advice or seek help from a specialist chemical waste disposal company.

3.8 SPECIAL PRECAUTIONS WITH CARCINOGENS ETC.

Table 3.2 lists some carcinogens that might be encountered in biomedical laboratories. Most work should be done in fume cupboards or safety cabinets of special design. Bench tops should be completely nonabsorbent otherwise substances may be absorbed and retained. Plastic aprons should always be worn over the ordinary overalls. They should be discarded for decontamination (if necessary) and laundering after each work session. Rubber or disposable gloves should be worn and washed well in cold running water before they are removed and discarded. Hands should be washed frequently - and particularly after a spillage - in cold running water.

3.8.1 Spillage

The above precautions (see Section 3.6) apply to carcinogens, but usually on a smaller scale. The most important precaution is against contact of the agent with any part of the body. Protective and other exposed clothing and shoes should be removed and the skin washed thoroughly in cold water before washing with soap and water. The clothing should be washed in cold water and then laundered.

3.8.2 Disposal

Disposal may be difficult and requires detailed knowledge of the composition of the substance. Chemical methods include

Kjeldahl digestion and dry combustion in silica tubes in a furnace. Ordinary incineration should not be attempted without professional advice as some carcinogens are volatile and will contaminate the lower atmosphere.

TABLE 3.3 Critical amounts of radioactive nuclides used in hospitals

Nuclide	Notification of use Bq	For controlled areas (internal)			Hazard assessment level Bq	Notification of incident Bq
		Air concentration $Bq.m^{-3}$	Surface contamination $Bq.cm^{-2}$	Minimum total activity Bq		
Cs137	5.10^5	6.10^2	5.10^2	4.10^7	2.10^{12}	2.10^{10}
Ca45	5.10^5	3.10^3	7.10^3	3.10^8	2.10^{14}	2.10^{10}
C14	5.10^5	1.10^4	1.10^4	9.10^8	2.10^{15}	2.10^{11}
Cr51	5.10^6	9.10^4	1.10^5	7.10^9	2.10^{15}	2.10^{12}
Co60	5.10^4	2.10^2	8.10^2	1.10^7	2.10^{12}	2.10^9
Cu64	5.10^6	9.10^4	5.10^4	4.10^9	2.10^{15}	2.10^{10}
Ga67	5.10^5	6.10^4	4.10^4	3.10^9	2.10^{16}	2.10^{12}
Au198	5.10^5	9.10^3	6.10^3	5.10^8	2.10^{15}	2.10^{10}
H3	5.10^6	2.10^5	4.10^6	3.10^{10}	2.10^{17}	2.10^{13}
In113m	5.10^6	6.10^5	2.10^5	2.10^{10}	2.10^{17}	2.10^{11}
I123	5.10^5	3.10^4	1.10^4	1.10^9	2.10^{15}	2.10^{12}
I125	5.10^4	3.10^2	1.10^2	1.10^7	2.10^{13}	2.10^{10}
I131	5.10^4	2.10^2	1.10^2	1.10^7	2.10^{13}	2.10^{10}
Ir192	5.10^5	9.10^2	5.10^3	8.10^7	2.10^{14}	2.10^{10}
Fe59	5.10^5	2.10^3	4.10^3	1.10^8	2.10^{13}	2.10^{10}
P32	5.10^5	2.10^3	2.10^3	1.10^8	2.10^{14}	2.10^{10}
K42	5.10^5	2.10^4	2.10^4	2.10^9	2.10^{15}	2.10^{10}
Ra226	5.10^3	3	8	2.10^5	2.10^{11}	2.10^8
Rb86	5.10^5	3.10^3	2.10^3	2.10^8	2.10^{15}	2.10^{10}
Se75	5.10^5	3.10^3	2.10^3	2.10^8	2.10^{13}	2.10^{11}
Na22	5.10^5	3.10^3	2.10^3	2.10^8	2.10^{15}	2.10^{10}
Na24	5.10^5	2.10^4	1.10^4	1.10^9	2.10^{15}	2.10^{10}
Sr85	5.10^5	6.10^3	1.10^4	6.10^8	2.10^{15}	2.10^{11}
Sr87m	5.10^6	6.10^5	1.10^5	1.10^{10}	2.10^{16}	2.10^{13}
Sr90	5.10^4	2.10	1.10^2	1.10^6	2.10^{12}	2.10^8
S35	5.10^6	9.10^3	2.10^4	8.10^8	2.10^{15}	2.10^{11}
Tc92m	5.10^6	6.10^5	4.10^5	3.10^{10}	2.10^{16}	2.10^{13}
Tl201	5.10^6	9.10^4	7.10^4	6.10^9	2.10^{16}	2.10^{12}
Xe133	5.10^6	6.10^6	N.A.	5.10^{11}	2.10^{20}	2.10^{14}
Y90	5.10^5	3.10^3	2.10^3	2.10^8	2.10^{14}	2.10^{10}

3.9 PRECAUTIONS WITH RADIOACTIVE SUBSTANCES

Some of the unsealed radioactive isotopes that may be used in clinical and biomedical laboratories are listed in Table 3.3. Those most commonly used are carbon 14, hydrogen 3 (tritium) and phosphorus 32, all of which emit beta particles, and iodine 125 which emits X-rays and gamma-rays. Iodine 125 is much more toxic than most of the other substances used because its compounds are volatile. Work with it should therefore be done in a fume cupboard but if the activity is less than 4 MBq (100 μCi) the hazard is minimal with good technique.

In clinical laboratories activities greater than 4 Mbq are unlikely to be encountered. The hazards in handling such compounds include skin contamination and the deposition of isotopes in the body, spread of contamination, e.g. as a result of spillage or careless handling or disposal, and external beta and gamma radiation. The units used in this kind of work are shown in Table 3.4.

TABLE 3.4 Units used in work with radioactive substances in bio-medical laboratories

---

Units of dose

Absorbed dose = energy absorbed/kg body weight

1 Gray	= 1 joule/kg = 100 rads
1 Sievert (Sv)	= 100 rems
1 milliSievert (mSv)	= $Sv/10^3$
1 microSievert (μSv)	= $Sv/10^6$

Activity units

1 Becquerel (Bq)	= 1 disintegration/second
1 kilo Becquerel (KBq)	= 10^3Bq
1 mega Becquerel (MBq)	= 10^6 Bq
37 MBq	= 1 millicurie (mCi)

As it is impossible to avoid all radiation (which naturally occurs at a rate of about 1mSv/year in this part of Europe) the exposure to radiation in a laboratory or medical investigation should be governed by the ALARA ('as low as readily achievable') principle. If a worker is likely to be exposed to more than 15 mSv/year then the laboratory in which he works must be designated a 'Controlled Area' to which access is strictly controlled. If the exposure is not likely to exceed 5 mSv/year then the laboratory is designated a 'Supervised Area'. Most clinical laboratories where isotopes are used will be designated Supervised Areas.

All hospitals and institutions where radioactive substances are used will have a Radiation Protection Officer (RPO) and local Codes of Practice as well as being subject to any national legislation.

3.9.1 The isotope laboratory

A separate room should be designated for work with radionuclides and the international radiation warning sign should be displayed on the door (see colour plate section). Bench surfaces should be impermeable (e.g. fitted with Formica). If they are not, local rules may permit them to be covered with a suitable disposable material, e.g. Benchkote (Gallenkamp).

Isotopes and kits containing isotopes should be stored in locked cupboards marked with the radiation warning sign. They should not be left on benches after use. Work should be done on trays. Instruments used should preferably be disposable.

A record should be kept of the quantities of radioactive substances purchased, used, stored and disposed of. Monitoring equipment should be available.

3.9.2 General precautions in the isotope laboratory

Only designated staff should work with radioactive substances. Protective overalls should be worn in, but not taken outside the radioactive area until they have been decontaminated. Rubber or disposable gloves should be worn when handling these substances and care taken to avoid touching (and therefore contaminating) any article or fitting outside the radioactive area. They should be washed while still on the hands and the hands should be washed after the gloves have been taken off. Discarded gloves and clothing should be placed in plastic bags to await disposal. Kits and radioactive materials should not be held close to the body. There should be a total ban on mouth pipetting, eating, drinking, smoking and the application of cosmetics. Advice on the wearing of film badges should be obtained from the RPO (they are not usually necessary for work with carbon 14, sulphur 35 and iodine 125) as well as the necessity for monitoring hands etc. before leaving the laboratory.

3.9.3 Personal contamination

If personal contamination occurs the RPO should be informed at once. The contaminated area should be washed well with cold water taking care not to spread the contamination, e.g. to the eyes. It should then be monitored and if the activity level is greater than 5 Bq/cm (or 50 Bq/cm in the case of hydrogen 3 and carbon 14) washing should continue with soap and tepid water until monitoring shows that the contamination has been reduced to below 5 Bq/cm.

3.9.4 Spillage

Local Codes of Practice will give instructions and the action to be taken will depend on the amount and activity of the substance spilled. Emergency kits should be provided. These should contain protective clothing and equipment similar to that described in Section 3.6.1 as well as monitoring equipment.

People leaving the contaminated area should be monitored to ensure that they are not carrying radioactive material on their clothing and shoes etc. Contaminated articles and waste should be placed in plastic bags pending removal for decontamination or disposal. Benches and contaminated areas should be washed with water, taking care not to spread the agent any further, until the activity is reduced to below 50 Bq/cm for hydrogen 3 and carbon 14 and 5 Bq/cm for other substances.

3.9.5 Disposal of radioactive waste

The amount of radioactive waste in clinical and biomedical laboratories is usuallly quite small. Disposal methods are usually written into local Codes of Practice.

The amount of radioactivity in the remains of kits should be calculated and recorded and the material may then usually be washed down the sink or sluice with large volumes of water. The solid remnants, packing, etc. should be autoclaved. The remains of scintillation fluids should be stored in labelled containers and removed by specialist contractors. This must not be flushed into the public sewers.

Small amounts of some other liquids containing long life substances may be flushed into the public sewer, subject to permission by the local authority and the HSE inspector. Similarly, flammable material may be burned if a suitable incinerator is available. Advice and permission from the local authority and HSE must be obtained for the disposal of solid, noncombustible radioactive waste, including ash from incinerators that have been used to dispose of volatile and flammable radioactive materials. Short-life materials may be allowed to decay naturally in a safe store.

3.9.6 Further information

The above is merely an outline. For more detailed information the reader is referred to the publications of the Committee of Vice Chancellors and Principals of the Universities (1966), WHO (1982) and Taylor (1985).

3.10 FUME CUPBOARDS

Fume cupboards are intended to remove toxic vapours and gases, and in some cases finely particulate material, from a work bench so that the operator does not inhale them. They are not places in which to store chemicals with unpleasant odours.

The British Standard Institution (BSI 1982) defines a fume cupboard as a 'partially enclosed workplace'. This is because it is essentially a ventilated box the front of which is usually a vertically sliding framed glass sash which can be opened so that the operator can work inside with his hands. Air is extracted through this opening and ducted to atmosphere.

3.10.1 Construction

Materials handled in fume cupboards may be highly corrosive so the carcass and trunking must be made of materials that are resistant to the chemicals that may be used in them. This is reflected in the cost of various models. The user should ascertain that his fume cupboard is constructed of suitable materials.

3.10.2 Air flow

The cupboard is ventilated by an extractor fan, usually placed on the roof of the building and connected by ducting to the top of the cupboard. Air is extracted through the open sash at (usually and perhaps ideally) between 0.6 and 0.9 m/s. The air velocity through the front depends on the size of the aperture, which in turn is controlled by the operator. To ensure that this velocity is always the same regardless of the position of the sash the best types of fume cupboards have balancing or ballasting mechanisms. The air flow should be monitored regularly (e.g. weekly) with an anemometer. Vane or thermistor anemometers suitable for measuring airflows in fume cupboards and microbiological safety cabinets are described by Collins (1988).

3.10.3 Siting

The air flow through a fume cupboard is influenced by its position in the laboratory in relation to the movement of air from doors, openable windows and mechanical ventilation ducts. A cross-draught may compromise the efficiency of the cabinet so that toxic vapours escape back into the room. The principles of siting fume cupboards are much the same as those for siting microbiological safety cabinets (Collins, 1988).

3.10.4 Effluents

Toxic and corrosive effluents must not be discharged to the lower atmosphere and no fume cupboard outlet should be near to openable windows or in places where the effluent can be blown to areas used by people. It may be necessary to erect a high chimney to discharge the effluent safely.

3.10.5 Standards and information

The British Standard Institution is still working on a standard but there is a useful Draft Document (BSI, 1982). Useful information is given by Hughes (1980). Methods for testing fume cupboards for their ability to capture and retain vapours and particles are being developed along the lines of those used for microbiological safety cabinets (Clark, 1983a).

3.11 HAZARDS OF EQUIPMENT USED IN CHEMICAL LABORATORIES

Some hazards arising from the use of glassware, analytical and electrical equipment are discussed by Kennedy (see Chapter 2), hazards associated with compressed gas cylinders by Stevens (see Chapter 3), fire hazards by Stevens (see Chapter 4) and infection hazards by Collins (see Chapter 6).

3.12 REFERENCES

Bretherick, L. (1979) *Handbook of Reactive Chemical Hazards*. 2nd edn. London: Butterworths.
Bretherick, L. (1981) *Hazards in the Chemical Laboratory*. London: The Royal Society of Chemistry.

BSI (1982) *Laboratory fume cupboards (DD80)*. London: British Standards Institution.

Cater, D.B. & Hartree, E. (1977) Carcinogens, mutagens and teratogens. In *Safety in Biological Laboratories*. (eds. E. Hartree and V. Booth). London: The Biochemical Society. pp. 47–54.

Chester Beattie Research Institute (1977) *Precautions for Laboratory Workers who Handle Carcinogenic Aromatic Amines*. London: National Cancer Institute.

Chivers, G.E. (1983) *The Disposal of Hazardous Wastes*. Occupational Hygiene Monograph No. 11. London: Science Reviews.

Chivers, G.E. (1985) Precautions in chemical laboratories. In *Handbook of Laboratory Safety and Health Measures* (ed. S.B. Pal). Lancaster: MTP Press. pp. 87–119.

Clark, R.P. (1983a) *The Performance, Installation, Testing and Limitations of Microbiological Safety Cabinets*. London & Leeds: Science Reviews.

Clark, R.P. (1983b) Review article: Formaldehyde in pathology departments. *Journal of Clinical Pathology* **36**, 839–846.

Collins, C.H. (1988) *Laboratory Acquired Infections*. 2nd edn. London: Butterworths.

Committee of Vice Chancellors and Principals of the Universities of the United Kingdom (1966) *Radiation Protection in Universities*. London: Association of Commonwealth Universities.

Cronor's *Substances Hazardous to Health*. (1986) (ed. A. Gardner Ward). New Malden: Cronor Publications.

DHSS (1976) Use of perchloric acid and treatment of equipment affected by perchloric acid. HN(76)95. London: Department of Health and Social Security.

HSE (1977) *Storage of Highly Flammable Liquids*. Guidance Note CS2. London: HMSO.

HSE (1981) *Code of Practice for work with Radioactive Substances*. London: Health and Safety Executive.

HSE (1984) *Occupational Exposure Limits*. Guidance Note EH40. London: Health and Safety Executive.

Hughes, D. (1980) *A Literature Survey and Design Study of Fume Cupboards and Fume Dispersal Systems*. Leeds: Science Reviews.

Kennedy, D.A. & Jennings, N. (1985) Ozonic ether. *Gazette, Institute of Medical Laboratory Sciences* **29**, 220

MCA (1972) *Laboratory Chemicals Disposal Manual*. Washington: Manufacturing Chemists Association.

MRC (1981) *Guidance for Work with Chemical Carcinogens*. London: Medical Research Council.

NIOSH (1983) *Registry of the Toxic Effects of Chemical Substances.* National Institute of Occupational Health and Safety. Washington: Government Printing Office.

NIOSH (1985) NIOSH recommendations for occupational safety and health standards. *Morbidity and Mortality Weekly Reports* **34**, 6S–13S.

Smith, J.H. (1985) Safety measures in a clinical chemistry laboratory. In *Handbook of Health and Safety Measures.* (ed. S.B. Pal). Lancaster: MTP Press. pp. 161–183.

Taylor, D.M. (1985) Radiation protection in radionuclide investigations. In *Handbook of Health and Safety Measures.* (ed. S.B. Pal). Lancaster: MTP Press. pp. 235–254

Warren, P. and Potts, M. (1985) *Dangerous Chemical Emergency Spillage Guide.* London: Walters Samson.

WHO (1982) *Protection against Ionising Radiation. A Survey of Current World Legislation.* Geneva: World Health Organisation.

4 COMPRESSED AND LIQUEFIED GASES

J.F. Stevens

Compressed and liquefied gases offer mechanical, fire, explosive and asphyxiation hazards.

4.1 COMPRESSED GASES

Compressed (and liquefied) gas cylinders are colour coded according to British Standard 349 (BSI, 1973 a, b) (Table 4.1). The valve connections of flammable gases have a left-hand thread; those of non-flammable gases have a right-hand thread. Oxygen cylinders have a right-hand thread.

TABLE 4.1 Colour codes of compressed gas cylinders used in biomedical laboratories

Gas	Colour code
Air	Grey, shoulder half black
Carbon dioxide	Black
Hydrogen	Red
Nitrogen	Grey
Oxygen	Black, shoulder white

For mixtures of hydrogen, nitrogen and carbon dioxide for anaerobic work see Section 2.9.2.1.

The local fire brigade should be informed that compressed gas cylinders are used. Fire officers will need to know which gases, where they are stored and used, and how many cylinders are on the site. All rooms and places where compressed and liquefied gases are used and stored should be labelled on both sides of the doors, with the international signs (see colour plate section), orange for flammable and green for non-flammable gases.

4.1.1 Storage

Gas cylinders not in use should be stored outside the laboratory in soundly constructed sheds or small buildings. These buildings should not be heated and the cylinders stored in them should be protected from strong sunlight. Electric light switches should be spark-proof or outside the building. The stores should be locked and be as vandal-proof as possible. Flammable and non-flammable gases should be stored in separate buildings or separate compartments of the same building. Ideally, oxygen cylinders should also be stored apart from other gases.

Valve protection caps should always be in place and not removed until the cylinder is ready for use. Empty cylinders should not be stored with full cylinders. Accidental connections between empty cylinders and pressurized systems can be hazardous.

4.1.2 Moving and securing cylinders

Gas cylinders are very heavy and can cause physical damage if they are handled carelessly or not properly secured. The valve may become detached, when the cylinder becomes an unguided missile, jet propelled at about 2500 p.s.i. There have been several reports of such incidents where there has been a great deal of damage (see Section 2.5).

Cylinders should not be moved without mechanical assistance. They should not be dragged along floors or handled by their valves. Purpose-made trolleys are available and these should always be used. In both stores and laboratories gas cylinders should be secured to walls or benches so that they cannot fall over. Free-standing cylinders are dangerous. Chains and purpose-made straps are available.

4.1.3 Compressed gas installations

In large and certainly in new laboratories the gas cylinders in current use should be outside the building, adequately protected from the weather and vandals, and the gas should be piped into the rooms where it is to be used. Such installations are the province of compressed gas engineers; town gas systems, which operate at lower pressures are inadequate. Piped systems are not always possible however, and in the interests of safety of staff and premises, only the cylinders in use should be in laboratory rooms.

4.1.4 Use of gas cylinders in the laboratory: reducing valves

Properly secured cylinders should be near the apparatus they supply but not near to radiators, sunlight or naked flames. (Even if the gas is non-flammable, escapes may extinguish such flames and if they are burning town gas or propane this may build up to an explosive concentration with air.)

Laboratory apparatus should never be connected to the main valve of a cylinder. Reducing valves must always be used. The outlets from these are often called 'side valves' to distinguish them from the 'main valves' which are integral parts of the cylinders. Reducing valves must be the correct type for the gas cylinders and although they are the property of the user, they should be serviced by the manufacturer in accordance with his recommendations. Unserviced reducing valves are hazardous (see Section 2.9.2.2). Spare reducing valves should be kept and these should also be serviced regularly, even if they have not been used. The main valve should always be opened before the side valve. The pressure gauges indicate the pressure in the cylinder and the pressure of the supply after reduction. Both valves should be closed when the gas supply is not required and certainly when the laboratory is unoccupied at nights and weekends etc.

Before removing and replacing a valve the user should ascertain whether it has a right-hand or a left-hand thread. It should not be necessary to use much force and the manufacturer's key or spanner, hand-held, should suffice. Hammers should not be used. The keys and spanners should be attached to cylinders or placed on hooks close to them. In some laboratories they are chained to the wall or bench. They should certainly not be left lying around on the bench.

The threads in the cylinder head and on the valve should be inspected and cleaned if necessary with a dry cloth. Oily or greasy materials should not be used to clean or lubricate the threads or there may be an explosion, especially with oxygen.

When the valve has been fitted to a full cylinder and before connections are made to any apparatus the main valve should be opened and the pressure noted. The side valve, pointed away from people or apparatus, should then be opened cautiously. Leaks may be obvious but if not they should be tested by brushing them with soap solution (not recommended for helium) or with a proprietory leak-detecting substance or apparatus.

Leaks may sometimes be rectified by tightening the connections, but again too much force should not be applied. If the valve connection to the cylinder leaks there may still be dirt in the threads and it may be necessary to remove the valve and clean them again. If the reducing valve leaks it should be returned to the manufacturer for inspection and repair.

If it is not possible to close the main valve, the cylinder with the reducing valve attached should be taken outside to a clear area and the pressure allowed to reduce slowly until the cylinder is empty. The manufacturer should be informed that the main valve is faulty.

4.2 SPECIAL GAS MIXTURES

Special mixtures of hydrogen, nitrogen and carbon dioxide are used for some equipment, e.g anaerobic cabinets. These are described in Section 2.9.2.1.

4.3 LIQUEFIED AND SOLIDIFIED GASES

4.3.1 Special precautions with oxygen, nitrogen and carbon dioxide

An escape of oxygen may enrich air and dramatically increase the risk of fire. Even an increase from the normal 21% to 25% will enhance combustion. Oxygen-rich air will even cause a cigarette or lighted pipe to burst into flames and will cause smouldering garments to burn fiercely.

Escapes of carbon dioxide and nitrogen (from cylinders or from liquid nitrogen or solid carbon dioxide) into confined spaces may depress the proportion of oxygen to a level that will not sustain life. People have been asphyxiated when they have entered such rooms. If the oxygen level is decreased to 5% life will not be sustained.

Carbon dioxide is in the solid phase in cylinders. Rapid discharge may cool the valves and tubes so much that ice forms and blocks them. The unrelieved pressure may be hazardous.

4.3.2 Liquefied petroleum gas (LPG)

LPG is used in laboratories for chemical apparatus or for bunsens. The cylinders must be stored vertically and not more than one cylinder should be in a laboratory room at any one time. These cylinders should not be near to radiators, naked flames, sparking electrical equipment or in direct sunlight. The correct valves must be used and serviced regularly.

In some installations the liquid gas is piped. Relief devices should be installed by the manufacturer's engineers.

Mistreated LPG cylinders are dangerous - they can behave like bombs and if one bursts the resulting explosion and fireball can do a great deal of damage.

4.4 ADVICE AND INFORMATION

This may be obtained from the DHSS Supplies Technology Division and from the booklet of British Oxygen Ltd (BOC

1980). Useful information about work with low temperature gases is given by the British Cryogenics Council (1970).

4.5 REFERENCES

BSI (1973a) BS349: *Identification of the contents of industrial gas cylinders.* London: British Standards Institution.

BSI (1973b) *Chart for the identification of the contents of industrial gas cylinders.* London: British Standards Institution

BOC (1980) *Safe under Pressure. Guidelines for all who work with cylinder gases.* London: British Oxygen Co.

British Cryogenics Council (1970) *Cryogenics Safety Manual: A guide to good practice.* Rugby: Royal Institute of Chemical Engineers.

5 FIRE IN THE LABORATORY

J.F. Stevens

Laboratories are 'high risk' fire areas because of the quantities and nature of flammable chemicals, solvents etc. used and stored in them. Ideally, only small amounts of flammable liquids should be held in the actual workplace. Bulk flammables should be stored elsewhere but this is not always practicable and if it is, is not always done. Chemistry and histopathology laboratories use more flammable chemicals and materials than other clinical and biomedical laboratories and are particularly prone to fires.

Fires in hospitals occur, unfortunately, quite frequently (in the UK, one every two or three days). Most are minor affairs, but the potential exists for major disasters, costing human lives and large capital losses. For this reason most health authorities include a contractual requirement for staff to attend fire lectures (and sometimes fire-fighting demonstrations) in their conditions of employment.

Most fires are preventable and training in sensible precautions is as important as training in dealing with fires. Some fires in laboratories arise from electrical faults (see Section 2.2.2) and laboratory managers should ensure that all electrical equipment conforms to the *Electrical Safety Code for Hospital Laboratory Equipment* (ESCHLE; DHSS, 1986). Others are the result of bad practices and poor maintenance and more than a few from negligence with naked flames.

5.1 CAUSES OF FIRES

There are many factors that predispose outbreaks of fire in laboratories. Apart from carelessness with smoking materials, two that are of particular importance are associated with electrical apparatus and the use of naked flames.

5.1.1 Common electrical faults responsible for fires

1. Overloading of circuits - too many appliances plugged into one socket outlet.
2. Long and badly disposed electrical cables - too near hotplates or similar equipment or where they are in contact with water.
3. Equipment left switched on unnecessarily, overnight or left unattended.
4. Equipment that uses flammables placed too near to naked flames.

5.1.2 Common naked flame faults associated with fires

1. Bunsens, full-flame or pilot, left unattended.
2. Bunsens close to containers of flammable liquids or materials (e.g. cotton wool, laboratory request forms).
3. Bunsens and gas-heated appliances (including those using propane and acetylene) connected with perished, incorrect or poor quality tubing.
4. Use of matches instead of mechanical lighters. Matches are not always extinguished when discarded.

5.2 LABORATORY FIRE FIGHTING EQUIPMENT

All laboratories should be equipped with a variety of fire extinguishers (see below), fire hoses that reach all parts of the premises, sand buckets and fire blankets. These should be placed at 'fire stations' on each floor and near each stair well.

Each laboratory room should have at least one (appropriate) extinguisher and a fire blanket.

The principle of a fire extinguisher is to deprive a fire of oxygen. Fire extinguishers are colour coded according to use (see colour plate section). Use of the wrong type, e.g. water on electrical fires, is hazardous. There are five kinds in common use:

- Water (red) driven by carbon dioxide generated internally.
- Carbon dioxide (black).
- Dry powder (blue).
- Foam (cream).
- BCF (green).

Table 5.1 shows which should be used for various kinds of fires.

TABLE 5.1 Types of fire extinguishers

Type	Colour code	Use for	Do not use for	Notes
Water[*]	Red	Paper, wood, fabric	Electrical, flammable liquids, burning metals	Messy
CO^2 powder[†]	Black	Flammable liquids and gases, electrical	Alkali metals, paper	No mess
Dry powder	Blue	Flammable liquids and gases, electrical		Messy
Foam	Cream	Flammable liquids	Electrical	Messy
BCF[§]	Green	Flammable liquids, electrical		No mess

[*] CO_2 driven
[†] force of jet may spread burning paper etc.
[§] ventilate room well after use

5.3 FIRE WARNING AND OTHER NOTICES

A simply worded notice, printed in large red letters, should be displayed by the door of each room. It should tell the occupants what to do in case of fire. An example is shown in Table 5.2.

NB. Long, wordy and badly worded instructions defeat their object. No-one reads them.

'Fire stations' should be prominently labelled (in red) and arrow signs pointing towards them placed at strategic points. Fire alarms should be very clearly marked (in red). Clear instructions on reporting fires should be placed on or near every telephone. Fire exits and escapes should be prominently marked (in green). Direction signs to fire exits (in green) should

TABLE 5.2 Example of fire notice

FIRE

IF YOU FIND A FIRE

1. Sound alarm. Telephone No...............
2. Use fire extinguisher if safe to do so.
3. Leave building by fire escape route.

IF YOU HEAR THE FIRE ALARM

1. Close windows.
2. Turn off gas and electricity in room.
3. Leave building by escape route.
4. Go at once to Assembly Point.

DO NOT

1. Collect clothing and valuables.
2. Use lifts.

YOUR FIRE ESCAPE ROUTE

Turn left. Go down end stairs. Go out of door on right.

YOUR ASSEMBLY POINT

North side of car park by bottle bank.

be placed at strategic points. Assembly points should be clearly and prominently marked.

5.4 ASSEMBLY POINTS

These should be chosen with care and be near enough to the building so that staff can easily reach them but not too near to place people in danger or to inconvenience the fire brigade. At least two people should be in charge of each assembly point (in case one is absent) and have lists of who should be present.

5.5 FIRE INSTRUCTIONS

Staff should receive regular instruction on what to do in case of fire. The following points should be emphasized:

1. Raise the alarm.
2. If possible close windows and doors.
3. Attempt to extinguish fire with nearest appropriate fire extinguisher only if fire is small.
4. Leave the building by the nearest route as indicated on the Fire Notice.
5. Do not stop to collect personal belongings from elsewhere.
6. Do not use lifts (elevators). They may become 'fire chimneys'.
7. Go to Assembly Point. This is most important because if anyone appears to be missing a fireman may have to risk his own life by attempting a rescue.
8. If exits are blocked by fire or anything else go to a window. Signal presence, stay by the window but not directly below it or injury may result from broken glass or the feet of firemen.
9. If there is a lot of smoke, keep close to the ground, where the air is freshest, and if possible crawl to safety.

5.6 PRACTICAL INSTRUCTION

Periodic lectures on fire prevention are essential and all members of the staff should have practical instruction on how to use a fire blanket if another person's clothes or hair are on fire.

If at all possible some, if not all members of the staff should have practical instruction in the use of fire extinguishers and should extinguish fires in sand pits etc. under the supervision of professional firefighters.

5.7 FIRE DRILLS

The object of a fire drill is to ensure that all staff know how to evacuate the building, to determine how long that takes and to underline the importance of the assembly points.

5.8 FIRE SAFETY INSPECTIONS

Inspections are best conducted on a regular basis by professional fire prevention officers. They include ensuring that fire exits are not blocked, that the keys provided actually fit the fire escape doors, inspecting fire extinguishers and fire blankets, testing the water supply to fire hoses and advising on notices.

5.9 FURTHER INFORMATION

Useful information on the causes, prevention and control of laboratory fires is given by Everett and Jenkins (1973), Everett and Hughes (1979) and Steere (1980).

5.10 REFERENCES

DHSS (1986) Electrical Safety Code for Hospital Laboratory Equipment, ESCHLE. *Health Equipment Information No. 158*. London: Department of Health and Social Security.

Everett, K. and Jenkins, E.W. (1973) *A Safety Handbook for Science Teachers*. London: Butterworths. pp. 30–44.

Everett, K. and Hughes, D. (1979) *A Guide to Laboratory Design*. London: Butterworths. pp. 31–46.

Steere, N.V. (1980) Physical, chemical and fire safety. In *Laboratory Safety; Theory and Practice*. (Eds. A.A. Fuscaldo, B.J. Erlick and B. Hindman). New York: Academic Press. pp. 4–27.

6 MICROBIOLOGICAL HAZARDS

C.H. Collins

The hazards of work with microbes are twofold: infection, which may lead to serious disease, even death, and allergy, which may result in serious disability. Of these, infection is the most likely and it must be remembered that it may go beyond the laboratory and affect the community.

In the context of this chapter the words 'microbes', and 'micro-organisms' include bacteria, viruses, microfungi, protozoa and helminths.

Laboratory-associated infections have a long and dismal history (Collins, 1984, 1988) but concern in the 1970s about the high incidence of tuberculosis and hepatitis in laboratory staff led to a better understanding of the problem and attempts to inform laboratory workers about the hazards of certain technical procedures. The Public Health Laboratory Service published the monograph *Prevention of Laboratory-acquired Infections* (Collins, Hartley and Pilsworth, 1974) and four years later the Departments of Health produced the *Code of Practice for the Prevention of Infection in Clinical Laboratories and Post-mortem Rooms*, generally known as the 'Howie Code' (DHSS, 1978). In 1980 clinical and biomedical laboratories were subjected to the *Health and Safety at Work etc. Act 1974*. The reduction in the numbers of infections reported since that time (Grist, 1981; Grist and Emslie, 1987) may, at least in part, be attributed to these measures. Such infections still occur however, and in the UK have included streptococcal disease and dysentery (Kurl, 1981; Ghosh, 1982). Laboratory-acquired hepatitis B, although now rare in the UK, must still be reckoned with and this, together with fears of acquired immune deficiency disease (AIDS) prevents many laboratory workers from returning to old and unsafe techniques. Lessons may also be learned from laboratory-associated infections in other countries, eg. typhoid fever in the USA (Blaser and Feldman, 1980) acquired from quality control specimens.

6.1 CODES OF PRACTICE AND GUIDELINES

In the UK, the Health and Safety Executive (HSE) works with the Howie Code (DHSS, 1978), parts of which have been amended by the Advisory Committee on Dangerous Pathogens (ACDP, 1984, 1986) and the Health Services Advisory Committee (HSAC, 1985, 1986). At this time (June 1987) there is now no single official publication covering microbiological safety in the UK.

In the USA the appropriate document is *Biosafety in Microbiological and Biomedical Laboratories* formulated by the Centers for Disease Control and National Institutes of Health (CDC/ NIH, 1984). Some European countries have their own codes or guidelines, or in common with many non-European countries, use the *Laboratory Biosafety Manual* of the World Health Organisation (WHO, 1983).

6.2 ROUTES OF INFECTION IN LABORATORIES

Laboratory infections may be acquired by routes other than those associated with the spread of disease in the community because the organisms have unusual opportunities for entering the body and are concentrated into larger 'doses' than might be encountered in nature. The routes include:

- Inhalation
- Ingestion
- Injection
- Entry through cuts and abrasions in the skin
- Entry through the conjunctivae

Table 6.1 gives examples of organisms that may infect laboratory workers by the different routes. Whether or not an infection occurs depends on a number of factors, including the portal of entry, the number of organisms entering the body and the immune status of the worker (see Section 7.3).

TABLE 6.1 Examples of organisms that may infect laboratory workers by different routes

---

BY INHALATION

Brucella; Coxiella burnetti; Coccidioides immitis; Mycobacterium tuberculosis; Histoplasma capsulatum; adenoviruses; equine encephalitis viruses; lymphocytic choriomeningitis virus.

BY INGESTION

Campylobacter; Salmonella, especially *S. typhi; Shigella;* enteroviruses.

THROUGH THE SKIN*

Blastomyces dermatitidis; dermatophytes; hepatitis B virus; *Leptospira; Staphylococcus; Streptococcus.*

---

* By injection or by contamination of small cuts and scratches.

6.3 THE PRINCIPLE OF MICROBIOLOGICAL SAFETY

The principle of avoiding or minimizing the hazards of laboratory-associated infections is containment: blocking the routes of infection by placing barriers around the organisms, around the worker, and around the laboratory (see Fig. 1.1). These barriers may be defined as follows:

6.3.1 Primary barriers, around the organisms

1. Good laboratory practice which minimizes the possibility of them escaping from their culture vessels. (Even with the best techniques accidents and escapes do occur).
2. Disinfectants that deal immediately with accidents and escapes that contaminate benches and equipment.

3. Safety cabinets which prevent the inhalation of organisms if they escape into the air.
4. Autoclaves and incinerators that kill the organisms when investigations are finished.

6.3.2 Secondary barriers around the worker

1. Protective clothing including gowns, gloves, safety spectacles and (rarely) masks
2. Personal hygiene such as handwashing
3. Immunization and medical care.

6.3.3 Tertiary barriers around the laboratory

1. Safe disposal of infectious waste.
2. Security and limited access of public.
3. Care of invitees.

The extent of these barriers depends on the nature, and to some extent the amount, of the micro-organisms that are handled in the laboratory. Those that are unlikely to cause human disease require little more than good laboratory practice (GLP) - more to protect the work than the worker. Those that cause very serious disease require very strict containment - highly effective barriers.

6.4 HAZARD AND CONTAINMENT CLASSIFICATIONS

It is now generally accepted that micro-organisms may be classified into four Risk or Hazard Groups according to their properties, and that for each group there is an appropriate Containment Level for which barriers (precautions) can be specified. The four-group classifications of the World Health Organisation (WHO, 1979, 1983), the United States Public Health Service (CDC/NIH, 1984) and the UK ACDP (1984) are essentially the same. The lists of micro-organisms etc. within the Groups differs from country to country because the incidence of pathogenic micro-organisms varies in different populations and geographical locations. An organism which merits a high level of containment in one country because it is rare in the community, may deserve lower categorization in another where it is endemic and laboratory workers are as much, if not more, at risk outside the laboratory than they are in it. For this reason the WHO offers no lists and recommends that member states compile their own according to local circumstances.

The UK classification of micro-organisms on the basis of hazard, formulated by the Advisory Committee of Dangerous Pathogens (ACDP, 1984) is shown in Table 6.2. There are no published lists of organisms in Hazard Group 1. Table 6.3 lists the relatively common organisms in Hazard Group 2, which are likely to be encountered in clinical laboratories. Table 6.4 lists the relatively common organisms in Hazard Group 3. The full

TABLE 6.2 The Hazard Group classification of micro-organisms[*]

Group 1

An organism that is most unlikely to cause human disease.

Group 2

An organism that may cause human disease and which might be a hazard to laboratory workers but is unlikely to spread in the community. Laboratory exposure rarely produces infection and effective prophylaxis or effective treatment are usually available.

Group 3

An organism that may cause severe human disease and present a serious hazard to laboratory workers. It may present a risk of spread in the community but there is usually effective prophylaxis or treatment available.

Group 4

An organism that causes severe human disease and is a serious hazard to laboratory workers. It may present a high risk of spread in the community and there is usually no effective prophylaxis or treatment.

---

[*] Advisory Committee on Dangerous Pathogens (1984).

lists are published by ACDP (1984) and are likely to be revised from time to time.

TABLE 6.3 Micro-organisms in Hazard Group 2 likely to be encountered in clinical laboratories

Bacteria	Fungi	Parasites	Viruses
Acinetobacter	Aspergillus	Ancylostoma*	Herpesvirus
Actinobacillus	Candida	Ascaris	hominis*
Actinomyces	Cryptococcus	Babesia	Influenza*
Arizona	Epidermophyton	Balantidium	Newcastle*
Bacillus cereus	Microsporum	Clonorchis	Measles*
Bordetella	Sporothrix	Dipetalonema	Mumps*
Campylobacter	Trichophyton	Diphyllobothrium	Poliovirus*
Chlamydia other		Dracunculus	Vesicular
than avian		Fasciola	stomatitis*
C. psittaci		Giardia	Rubella*
Clostridium*		Hymenolepis	Louping ill*
Corynebacteria*		Loa	Tick-borne
Edwardsiella		Necator*	encephalitis*
Eikenella		Opistorchis	
Enterobacter		Paragonimus	
Erysipelothrix		Plasmodium	
Escherichia*		Pneumocystis	
Flavobacterium		Schistosoma*	
Haemophilus		Strongylides	
Klebsiella		Taenia	
Legionella		Toxocara	
Leptospira*		Trichinella	
Listeria		Trichomonas	
Moraxella		Trichiuris	
Mycobacterium		Trypanosoma other	
chelonei		than T. cruzi	
fortuitum		Onchocerca	
marinum		Wucheria	
ulcerans			
Mycoplasma			
Neisseria			
Nocardia			
Pasteurella			
Plesiomonas			
Proteus			
Providencia			
Pseudomonas other			
than P. mallei			
and P. pseudomallei			
Salmonella other			
than S. typhi			
and S. paratyphi			

continued

TABLE 6.3 *continued*
Serratia
Shigella
Staphylococcus
Streptobacilli
Streptococcus
Treponema[*]
Vibrio
Yersinia other than
    Y. pestis

[*] See ACDP (1984) for special conditions, e.g. vaccination, safety cabinets, gloves.

TABLE 6.4 Micro-organisms in Hazard Group 3 likely to be encountered in clinical laboratories

Bacteria	Fungi	Parasites	Viruses
Bacillus anthracis[*]	Histoplasma	Echinococcus[*]	Eastern equine
Brucella	Paracoccidioides	Leishmania	encephalitis[*]
Chlamydia psittaci	brazilensis	(mammalian)[*]	Hantaan fever
Coxiella burnetti		Toxoplasma gondii[*]	Hepatitis B
Francisella			Herpesvirus simiae[*]
tularensis[*]			Human T-cell
Mycobacterium			leukaemia
africanum[*]			viruses
bovis[*]			Japanese B
intracellulare[†]			encephalitis
kansasi			Korean
leprae			haemorrhagic
malmoense			fever
simiae			Kumlinge fever
szulgai			Lymphocytic
tuberculosis[*]			choriomenin-
Pseudomonas mallei			gitis
Pseudomonas			Murray Valley
pseudomallei			fever
Rickettsia			Powassa fever
Salmonella			Rabies[*]
paratyphi A			Rift Valley fever[*]
typhi[*]			St Louis
Yersinia pestis[*]			encephalitis
			Venezuelan
			equine
			encephalitis[*]
			Western equine
			encephalitis[*]
			Yellow fever[*]

[*] See ACDP (1984) for special conditions, e.g. vaccination, safety cabinets, gloves.
[†] Includes *M. avium* and *M. scrofulaceum*.

6.5 HAZARD GROUPS AND CONTAINMENT LEVELS IN CLINICAL AND BIOMEDICAL LABORATORIES

The majority of the organisms and viruses that are investigated in clinical and biomedical laboratories are in Hazard Groups 2 and 3 and therefore Containment Levels 2 and 3 apply (see Table 6.2). A few organisms, mostly normal flora, will be in Hazard Group 1, but for these Containment Level 2 can apply: micro-organisms may be handled at higher containment levels than those corresponding to their hazard groups. Hazard Group 4 contains viruses that are not investigated in clinical laboratories; work with them requires authority from the ACDP and HSE and the special conditions of Containment Level 4. Few such laboratories exist. Hazard Group 4 is not considered in this chapter.

The conditions for work at Containment Levels 2 and 3 (ACDP, 1984) include accommodation, ventilation, use of safety cabinets, provision of certain other equipment, specific techniques and personal protection (e.g. wearing gowns and gloves).

The recommendations and precautions described below are consistent with those in the Howie Code and the various publications of the ACDP and the HSAC. They are based on earlier works, however, in the USA (e.g. Phillips, 1961; CDC, 1974; NIH, 1974), and in the UK (e.g. Darlow, 1969, 1972; Collins et al., 1974). As with the various Codes and other documents, it is not possible to give the reasons behind some of the recommendations. They may be found, however, in the above publications and in that of Collins (1988).

6.6 THE LABORATORY AND SERVICES

6.6.1 Accommodation

The minimum requirements are those of the ACDP (1984) for Containment Level 2. They are similar to those set out in Section 1.5.2. It is not proposed to repeat them here in detail but essential points are a decent working environment, consistent with the requirements of the various Acts and Regulations now incorporated into the *Health and Safety at Work etc. Act 1974*; the provision of hand basins for washing hands (laboratory sinks will not do: see below); pegs or hangers for laboratory clothing; convenient and conveniently placed lockers for personal clothing and belongings, not in the laboratory rooms; convenient toilet facilities; staff rest rooms where food and drink may be consumed.

Where Hazard Group 3 organisms are handled there should be a separate 'Containment Laboratory' to the standard of the Containment Level 3 specifications of the ACDP. This room or suite should be locked when not in use. The doors and, if the local fire brigade so wish, the windows, should be labelled with the international Biohazard sign (see colour plate section).

6.6.2 Ventilation

There should be at least six air changes per hour in each room. If the ventilation is mechanical the air should flow from the corridors and offices ('clean areas') to the laboratories ('dirty areas') and thence to the atmosphere or for recirculation after treatment. This should prevent the dispersion of infectious air-borne particles from laboratories. More specific directional air flow is required for Level 3 work, where the air should be extracted to the atmosphere through high efficiency particulate air (HEPA) filters in microbiological safety cabinets.

6.6.3 Sterilization facilities

There should be at least one autoclave, specifically designed for laboratory use and dedicated to the decontamination of infected material. There should be a separate autoclave for 'clean' items, e.g. culture media.

There should be an incinerator on site and this should be under the control of the laboratory staff when it is used for the destruction of laboratory waste. If such an incinerator is not on site all infectious material should be autoclaved before it is taken out of the laboratory.

6.6.4 Access

There should be no public access to laboratory rooms. Access by other hospital/institute staff should be severely restricted. Maintenance workers should be supervised.

6.6.5 Handwashing facilities

At least one handbasin should be provided in each room as near to the door as possible. Larger rooms where several people work, or which have two doors, should have more than one such basin. Tablet soap should be provided rather than soap dispensers (these may become contaminated and may be emptied without visual warning). Paper, not cloth towels, should be provided.

6.6.6 Cleaning and domestic staff

These include workers in preparation ('wash-up') rooms as well as those who clean the premises. They should be informed about, but not frightened by the potential hazards that will surround them. The Howie Code (DHSS, 1978) contains Model Rules for these workers and these rules are reproduced in Table 6.5. Copies should be given to all members of the domestic staff - including contract cleaners - and enlarged versions prominently displayed where they are likely to be seen, not only in preparation rooms, but in domestic staff rooms and 'broom cupboards'.

TABLE 6.5 Model rules for laboratory domestic staff

---

1. Always wear the overall provided for your protection and see that it is properly fastened. Keep it apart from your outdoor clothing, not in your locker. Pegs are provided. Do not take your overalls home to wash.

continued

TABLE 6.5 *continued*

2. Do not wear your overall in the staff room or canteen. Take it off when you leave the laboratory to visit another part of the hospital.
3. Wash your hands often and always before leaving the laboratory or going to the staff room for food and drink or for a smoke. Cover cuts and grazes with waterproof dressings.
4. Do not eat, drink, smoke, or apply cosmetics in any laboratory. Use the staff room.
5. Do not touch any bottles, tubes or dishes on any of the laboratory benches unless you have been told by the Safety Officer or your Supervisor that it is safe for you to do so.
6. Do not dust or clean any work benches without permission from one of the laboratory staff.
7. If you have an accident of any kind, or knock over or break any bottle, jar, or tube, or piece of equipment, tell the Safety Officer or your Supervisor or one of the laboratory staff at once.
8. Do not attempt to clear up after any accident without permission from a senior member of the laboratory staff. Do not pick up broken glass with your fingers. Use a dust pan and brush. Follow instructions of senior members of staff.
9. Do not enter any room which has the red and yellow Danger of Infection sign on the door until the occupier tells you that it is safe to do so.
10. Do not empty any laboratory discard containers unless a label or an instruction say that you may do so.

IF YOU WORK IN THE WASH-UP ROOM FOLLOW THESE INSTRUCTIONS AS WELL AS THOSE ABOVE:

1. Do not handle or wash any material that comes from the laboratory until it has been sterilised (autoclaved) or one of the laboratory staff or your Supervisor has told you that it is safe.
2. Do not place broken glass in plastic disposal bags. Use the labelled containers provided.
3. Do not work with the autoclave until you have been taught how to do so by your Supervisor and the Safety Officer is satisfied that you are competent to operate it. Follow the written instructions displayed near to it at all times.
   If you cut or prick yourself or have any accident which injures you, however slightly, report it to your Supervisor *at once* and see that the Safety Officer records it in the Accident Book. This may save you a lot of trouble later.

If you obey these simple rules you will be as safe as anyone else who works in the hospital, BUT if you are ill tell your doctor where you work and ask him to talk to one of the doctors in the laboratory. If in doubt about anything, ask the Safety Officer.

Reproduced from the *Code of Practice for the Prevention of Infection in Clinical Laboratories* (DHSS, 1978) by permission of Her Majesty's Stationery Office.

6.7 SPECIMEN TRANSPORT AND RECEPTION IN CLINICAL LABORATORIES

6.7.1 Specimen containers

These should be robust and should not leak according to the tests prescribed by the British Standards Institution (BSI, 1972, 1975) or otherwise be acceptable to the HSE. It is well known that certain types of 'plug in' and 'snap closure' containers release small amounts of their contents when they are opened. Screw-capped containers are safer but even they may leak if they are poorly made or, with re-usables, the washers are defective, which may happen after they have been washed and re-used several times. Disposable containers should be purchased only from reliable suppliers; re-usables should be carefully inspected before reissue.

6.7.2 Plastic bags for specimen containers

In some hospitals all specimens in their containers are placed in self-sealing plastic bags. In others these bags are used only for containers with specimens from patients known or suspected of suffering from hepatitis, AIDS, other Hazard Group 3 infections or others thought to offer a risk of infection. Only bags designed for this purpose should be used. Bags should not be sealed by heat necessitating the use of a sharp instrument to open them, which could be hazardous.

6.7.3 Labelling and request forms

Self-adhesive labels should be used. Additional labels are recommended for 'Danger of Infection' specimens, e.g. on specimens from patients with hepatitis, AIDS or Hazard Group 3 infections (see Section 6.9 and colour plate section).

Request forms should not be wrapped around specimen containers, nor placed with them in the same plastic bags. Some makes of plastic bags have separate pockets for forms.

6.7.4 Transport within hospitals

6.7.4.1 *Laboratory porters and messengers*   The Howie Code includes Model Rules for those who carry specimens. These have been revised (HSAC, 1986) and are shown in Table 6.6. Copies should be given to each member of the staff concerned and displayed at specimen collection points. The routes followed by porters and others who collect specimens from wards and departments should not include canteens or kitchens.

6.7.4.2 *Specimens brought by hand*   Within hospital premises these specimens should be in trays or boxes and conveyed in an upright position to minimize leakage. They should not be carried in pockets, sacks or bags. Several kinds of trays and boxes are available commercially: some, designed for kitchen and picnic use, are quite adequate and are inexpensive. The boxes and trays should be plastic or metal and easily cleaned. They should be disinfected at least once each week and immediately if a specimen has leaked.

TABLE 6.6 Model rules for laboratory porters and messengers

Some of your work in the hospital may involve handling material that contains germs capable of causing illness. You are not required to touch anything known to be infected but may do so accidentally and carry germs to your family and friends. These rules are made to protect both you and them.

1.   Wear your overall, properly fastened, whenever you are carrying specimens. Keep it apart from your outdoor clothing, not in your locker. Pegs are provided. Never wear your overall in the staff room or canteen. If you do you could spread infection.
2.   Cover any cuts or grazes on your hands with a waterproof dressing.
3.   Carry all specimens in the trays or boxes provided, not in your hands or in your pockets. Touch specimen containers as little as possible.
4.   If you do touch them wash your hands as soon as possible afterwards.
5.   Wash your hands often, especially before meal breaks and at the end of a spell of duty.

continued

TABLE 6.6 *continued*

6. If a specimen leaks into the tray or box tell the laboratory reception staff and ask them to get it made safe.
7. If you drop and break a specimen do not touch it or try to clear up the mess. Stay with the specimen to prevent other people touching it and send someone to the laboratory for help.
8. If you drive a van make sure that you have a bottle of hospital disinfectant and some cotton wool with you. If a specimen leaks in your van and runs out of the tray or box, pour disinfectant over the mess and cover it with cotton wool. Do not mop it up. Drive to the laboratory for help.
9. If you have a breakdown or accident do not let anyone touch the specimens unless they come from a hospital and know the rules.
10. Handle specimens packed in boxes gently at all times.
11. Take care when carrying waste or rubbish from the laboratory to the incinerator or rubbish tip. There may be broken glass or needles. If you find these tell the Safety Officer. Special containers are provided for glass, syringes and hypodermic needles.
12. If you cut or prick yourself or have an accident, however small, tell the laboratory Safety Officer and see that he enters the facts in the Accident Book. This may save you trouble later.
13. Never eat, drink or smoke when you are carrying specimens, or when you are in any laboratory room.
14. If you are ill tell your doctor your place of work in case you have caught a germ from the laboratory. Ask him to talk to one of the laboratory doctors.

If you obey these simple rules you will be as safe as anyone else who works in the hospital. Do not be afraid to ask for advice from the laboratory Safety Officer – that is why he is there.

Reproduced from *Safety in Health Service Laboratories: the labelling, transport and reception of specimens.* Health and Safety Executive (1986) by permission of Her Majesty's Stationery Office.

6.7.5 Transport between hospitals

If specimens are carried any distance, over the public highway, by hand or in any form of transport then they should be in closed and leak-proof boxes which are usually made of metal or strong plastic. Such trays and boxes are available commercially. They should carry a label stating that they should not be opened except in a laboratory and the address and telephone number of the laboratory should be printed clearly on that label.

6.7.6 Specimens through the post

Specimens sent through the post should conform to the Post Office regulations for pathological material (see the *Post Office Guide*, obtainable at Crown Post Offices). The essential features are that the primary container - that which holds the specimen - is robust, not easily broken and does not leak (but leakage depends very much on the user!) and that this is packed firmly in the outer container with absorbent material in case of leakage. The outer container is usually of strong card or plastic.

The most suitable outer containers for sending specimens through the post are probably those already approved by the DHSS and supplied to NHS hospitals. Other containers are available commercially but before they are purchased Post Office approval should be sought (address in the Post Office Guide).

6.7.7 Reception at the laboratory

6.7.7.1 *Specimen reception rooms*   These should be furnished as laboratories rather than as offices (offices should be in separate rooms). The same type of flooring, handbasins, benches with impervious and easily cleaned surfaces are essential. There should be a supply of suitable disinfectant (freshly prepared according to local custom: aged disinfectant solutions are usually ineffective).

6.7.7.2 *Reception staff*   These are usually recruited from offices and have no knowledge of the potential hazards of handling specimens. Model Rules (DHSS, 1978, revised by HSAC, 1986) have been published and are reproduced in Table 6.7. Copies of these rules should be given to all members of the staff, and larger printed versions posted in the offices.

The reception staff will need to know how to use a disinfectant in an emergency and when to call upon a member of the laboratory staff (usually the designated safety officer) to deal with a spillage or a leaking or broken specimen. They should not attempt to do this themselves. They should also know whom they should call if one of their number has been injured - however minor the injury may appear to be.

TABLE 6.7 Model rules for laboratory reception staff

Much of the work of laboratory reception staff will involve handling containers and packages which are being used to send infectious samples to the laboratory. You are not required to touch anything known to be infected but you may do so accidentally and could carry germs to your family or friends. These rules are made to protect both you and them.

1. Wear your laboratory overall, properly fastened, at all times in the reception room and when visiting laboratories. Keep it apart from your outdoor clothing, not in your locker. Pegs are provided.
2. Never wear your laboratory overall in the staff room, canteen or dining room. If you do you may spread infection.
3. Wash your hands often during work and always before you leave the reception room. Cover cuts and grazes with waterproof dressings.
4. Never eat, drink, smoke or apply cosmetics in the reception room. You may infect yourself. Go to the staff room.
5. Never lick labels.
6. If a leaking or broken specimen arrives do not touch it or any others in the same box or tray. Ask a member of the medical or scientific staff to deal with it.
7. Do not unpack or remove from its plastic bag any specimen with a label indicating a danger of infection. These are delivered in this way because there is a risk of hepatitis and other diseases. They must be delivered unopened directly to the relevant department.
8. Keep all the specimens together on the reception bench. Never put them on your desk or anywhere else.
9. Twice each day, eg. before lunch and when you finish work for the day, wash down the specimen bench with the disinfectant and disposable cloths provided.
10. Do not allow visitors to touch anything on the specimen bench. Keep children out of the reception room. They do not know the rules and may become infected.

If you obey these simple rules you will be as safe as anyone else who works in the hospital, BUT if you are ill tell your doctor where you work and ask him to talk to one of the doctors in the laboratory.

Reproduced from *Safety in Health Service Laboratories: the labelling, transport and reception of specimens*. Health and Safety Executive (1986) by permission of Her Majesty's Stationery Office.

6.8 GENERAL LABORATORY PRECAUTIONS AT CONTAINMENT LEVEL 2

6.8.1 Avoiding the inhalation of micro-organisms

Aerosols – fine suspensions in air of minute droplets of fluid – are formed whenever liquid surfaces are disrupted or liquids are treated violently. The smaller droplets evaporate rapidly, leaving any micro-organisms they contained as airborne infectious particles. These sediment very slowly and may be moved about by even the smallest of air currents. Ventilation systems may carry them to other rooms. If particles less than 0.5 microns in diameter are inhaled they are not removed in the upper respiratory tract, may enter the lungs and initiate an infection.

Larger aerosol droplets and infectious airborne particles behave in a different way. The droplets dry less rapidly and both droplets and infectious particles sediment rapidly and contaminate hands and benches etc. (see Section 6.8.4). Infectious particles may also be released into the air when dried cultures or cultures of sporulating fungi are opened.

Examples of how aerosols and infectious airborne particles are produced, and relevent precautions are:

Inoculating loops   Loops larger than 3 mm in diameter or incompletely closed will shed their contents.

Vigorous spreading of culture plates and slides with loops Gentle movements are safer.

Flaming charged loops in ordinary bunsens   Hooded bunsens, or better still, disposable plastic loops should be used.

Pipetting   Blowing out the last drop will cause bubbles to form at the tip and these will burst. Pipettes should be drained, not blown out.

Vigorous mixing of cultures   Sucking and blowing with pipettes will produce bubbles and release clouds of aerosols. Vortex and other mixers are better but culture tubes or bottles should be stoppered.

Falling drops   Drops sometimes fall from the tips of pipettes. When they hit a hard surface aerosols and small droplets are thrown up. An absorbent bench covering, preferably soaked in disinfectant reduces aerosol formation.

Centrifugation   Fluids may fly out of unstoppered centrifuge tubes when the machine is started or stopped. Impact with the bowl of the centrifuge creates aerosols which are sprayed into the room even if the lid is closed. A line of fluid or dried material may also be visible inside the bowl. Screw capped tubes and sealed buckets should be used. Breakage of tubes releases massive aerosols. Sealed centrifuge buckets contain them (and infected glass fragments).

Homogenization   Aerosols form in the container and the temperature rises during operation. Opening the container releases them. This does not happen with stomachers.

Mixing in unstoppered containers   Bottles and tubes should be stoppered. Vortex mixers are safest.

Dropping cultures   There are no obvious precautions but cultures in plastic petri dishes release less aerosols than glass ones when dropped.

Probes of automated equipment   These often move abruptly, generating aerosols and splashes. Guards should be fitted and the movement slowed down.

Pouring   When one fluid is poured into another a Rayleigh jet is formed; small bubbles and aerosols are generated. Infected fluids should be poured through a funnel which is sitting on a disinfectant pot with its tip beneath the surface of the fluid.

6.8.2 Avoiding the ingestion of micro-organisms

Organisms may be taken into the mouth or transferred to the mouth from hands and fingers that have been contaminated directly or indirectly by accidental spillage, the outside of specimens containers and large-droplet aerosol formation (see Section 6.8.1).

Mouth pipetting This is the cause of many enteric infections. Mouth pipetting, even of sterile fluids should be banned. Pipetting devices should be provided and as the safest for any purpose is that which best suits the individual worker, it may be necessary to purchase several different types. In-house training in the use of these devices, and of rubber teats (Collins and Lyne, 1984) is desirable and helps to overcome objections to their use.

Infected hands Hand washing should be encouraged to break the hand-to-mouth route. Hands are often unwittingly contaminated from the outsides of specimen containers, cultures or equipment.

Eating, drinking and storing of food This should be banned in the laboratory: contamination is too easy.

Smoking Organisms may be transferred from contaminated fingers to the mouth. It should be banned.

6.8.3 Avoiding injection of micro-organisms

The skin may be penetrated by sharp instruments and glass. Many infections have been caused by hypodermic needles (Collins & Kennedy, 1987).

Hypodermic needles and syringes These should not be used in place of pipettes. There are instruments for removing septum caps so that pipettes may be used. Exceptionally needles may be replaced by cannulas, but these are often quite sharp.

Needle probes of automated equipment These take samples from septum capped bottles. This is a newly discovered hazard. Shields should be fitted where possible.

Glass Pasteur pipettes   Hands are easily stabbed. Soft plastic Pasteur pipettes are safer.

Poor quality culture tubes   These may break when they are capped or a stopper is pushed in. The operator may be 'inoculated'. Such glassware is hazardous and not economical.

Chipped culture tubes   Rims are usually contaminated. If a rim is also chipped it can 'inoculate' the operator. There should be a glassware inspection routine.

6.8.4 Avoiding infection through the skin

Micro-organisms can enter the skin through obvious and in-apparent cuts and abrasions on exposed skin, e.g. of hands and face.

All obvious cuts and abrasions should be covered with waterproof dressings. Hands should be washed frequently and not scrubbed. Barrier creams may be useful. Disposable gloves should be supplied. Their use depends on the nature of the organisms, state of the worker's hands, and personal choice.

6.8.5 Avoiding infection through the eyes

Eye infections may arise from splashes or from rubbing the eyes with infected fingers. Wearing safety spectacles should be considered.

6.9 SPECIAL PRECAUTIONS WITH HAZARD GROUP 3 BACTERIA AND VIRUSES

Any precautions mentioned below are in addition to those detailed above.

In clinical laboratories the two bacteria in Hazard Group 3 that are most likely to be encountered are *Salmonella typhi* and *Mycobacterium tuberculosis* (which includes *M. bovis* and *M. africanum*). The two viruses of general concern are those of hepatitis B (HBV) and AIDS (HIV, formerly LAV/HTLV-III). Laboratory precautions are related to the ways in which these agents may enter the human body (see Table 6.1) but in general are those specified by the ACDP (1984) for Containment Level 3 work.

The obvious problem is that the sender of a specimen may not have made a diagnosis and until the laboratory examination is complete, or nearly so, the identity of the organisms present is unknown and the laboratory worker does not know what the appropriate precautions are.

The general feeling among senior clinical microbiologists is that sputum is quite likely to contain tubercle bacilli, whatever the diagnosis and therefore it should be handled under Containment Level 3 conditions. Very few faeces specimens, on the other hand, are likely to contain *S. typhi* and may be safely handled under Containment Level 2 conditions. If, however, at some point in the investigation it is suspected that an organism that has been isolated is *S.typhi* then all the cultures and work should be transferred to a Containment Level 3 room. Subsequent specimens from the same source or same outbreak should be examined in that room.

This principle may reasonably be applied to any other material which might contain a Hazard Group 3 organism. When automated equipment is used (e.g. in chemistry and haematology) specially labelled and suspect material should be processed in batches, if possible, after the other specimens have been tested.

6.10 SPECIAL PRECAUTIONS FOR *SALMONELLA TYPHI*

It is reasonable to expect that laboratory workers who handle faeces specimens have received typhoid fever vaccine (see Section 7.3.2). A number of laboratory workers, however, including some who have been vaccinated, have contracted typhoid fever. In recent years some have acquired the infection from quality control specimens (Blaser and Feldman 1980).

The usual route of infection is oral but it is possible for the organisms to enter the body through the eye (hand-to-eye route) or the lungs, if, e.g. an aerosol is produced during slide agglutination tests. The most important precaution against laboratory-acquired typhoid fever is good hand hygiene: care not to get the organisms on the hands and frequent and thorough hand washing. The face and eyes should not be touched, nor should handkerchiefs.

Slide agglutinations and any work involving pipetting and transferring liquid cultures or suspensions of *S. typhi* should be done in a microbiological safety cabinet.

6.11 SPECIAL PRECAUTIONS WITH *MYCOBACTERIUM TUBERCULOSIS*

Laboratory workers who handle tubercle bacilli should have had BCG vaccine (see Section 7.3.1). The work should be done in a Containment Level 3 room. The outsides of specimen containers may be contaminated (Allen and Darrell, 1983) and precautions should be taken against hand to mouth contamination.

All procedures with sputum, suspected tuberculous material, and all manipulations with cultures of tubercle bacilli should be done in a microbiological safety cabinet.

Plastic disposable loops are safer than wire loops. They should be discarded into phenolic disinfectant.

Pathological material for microscopic examination should be spread on the slide with gentle movements to avoid splashing. Films of cultures should be made in small drops of saturated mercuric chloride with gentle movements. All slides should be dried in air in the cabinet before they are fixed by flaming in the usual way. They are then reasonably safe if the films are not touched by hand or allowed to come into contact with anything else. Tubercle bacilli may survive in heat-fixed films made in water or saline (Allen, 1981). Homogenizing and decontamination procedures should be done in stoppered tubes in a safety cabinet. Some of these procedures are safer than others. For methods see Collins and Lyne (1984), Collins, Grange and Yates (1986). Sealed buckets should be used for centrifuging and plastic, not glass pipettes should be used.

6.12 SPECIAL PRECAUTIONS WITH HEPATITIS B AND AIDS MATERIALS

Hepatitis B and AIDS are considered together because the precautions to be observed are very similar. Separate recommendations have been made (HSAC, 1985 for hepatitis and ACDP, 1986 for HIV), but there is a strong case for these to be unified (Hospital Infection Society, 1985) and there is no scientific reason for not doing so. The laboratory precautions are very similar, although laboratory workers are clearly more at risk from hepatitis B than from HIV. According to the Centers for Disease Control (Atlanta) as of August 15, 1985 no cases of AIDS that meet their definition (CDC, 1985) have been reported in laboratory workers (CDC, 1986)*.

6.12.1 Mode of transmission in the laboratory

It is generally accepted that in the UK between 1 in 800 and 1 in 1000 blood samples may contain the antigen HBsAg but only about 1 in 5 of these contain the 'e' antigen in the absence of antibody. It is these samples that offer a hazard to the laboratory worker. Hepatitis B is known to be have been transmitted to laboratory workers by ingestion and injection: i.e. by contaminated fingers and hand-to-mouth route, mouth pipetting, and cuts and pricks from contaminated glassware and equipment. There is no scientifically-based evidence of transmission by the air-borne route. It follows that Good Laboratory Practice, i.e. adherence to simple and sensible rules, will protect the worker from infection.

HIV is not known to have been transmitted to laboratory workers* and the infectivity seems to be low: according to Levy *et al.* (1985) 10^{13} hepatitis B particles may be present in 1 ml of blood, against only 10^4 of HIV. The restricted routes of infection of this virus suggest that precautions taken against hepatitis B are likely to prevent laboratory-acquired infection with HIV.

The precautions against infection described below are consonant with those of the HSAC (1985), the Hospital Infection Society (1985) and the ACDP (1986), and various communications from CDC.

---

* There has been one case in the US in which a laboratory worker was exposed to a large concentration of the virus (Press release, National Institutes of Health, November 1987).

6.12.2 Collection, labelling and transport of specimens

Specimens, especially blood, should be collected only by experienced staff. Precautions against needlestick accidents are particularly important (see Collins and Kennedy, 1987). HSAC (1985) states that 'needles must be removed from syringes before blood is discharged into the specimen container' but this is a recipe for contaminated fingers as well as needlestick. This requirement can be satisfied safely, however, if the needles are removed with needle forceps. The obvious alternative is to use a vacuum collection method.

Robust discard bins for needles and syringes (e.g. 'Burn Bins' or 'Cin Bins') should be provided where they are needed. They should not be overfilled.

The labelling and transport methods described in Section 6.7 are adequate. Suitable labels are available commercially (see colour plate section).

Reception room staff should not unpack or remove from their plastic bags any specimen with a 'Hepatitis Risk', 'AIDS Risk' or A2 'Danger of Infection' label.

6.12.3 Containment

6.12.3.1 *Accommodation*   According to HSAC (1985) testing for the presence of hepatitis B antigens and antibodies must be done in a separate room, either in a microbiological safety cabinet or one that conforms to Containment Level 3 requirements. In either case the room must not be used simultaneously for other work. Alternatively a Level 3 room without a continuous air flow or a safety cabinet may be used if only hepatitis testing is done there.

Full Level 3 conditions are required if concentrated virus is used for research purposes.

According to ACDP (1986) clinical laboratory work not involving the propagation or concentration of HIV requires containment conditions not less than those specified for Level 2 supplemented by specified rules (these are summarized below). Such work need not be done in a microbiological safety cabinet unless other (Hazard Group 3) pathogens are believed to be present.

Propagation and concentration of the virus requires full Level 3 containment. Clearly, these different provisions need harmonizing so that a laboratory handling both hepatitis B and AIDS material can use the same accommodation. HSE inspectors have been helpful here.

6.12.3.2 *Protective clothing* Dowsett–Heggie type overalls, plastic aprons, disposable gloves and safety spectacles are recommended.

6.12.3.3 *Glass and 'sharps'* Plastic should replace glass wherever possible. If glass must be used then only the very tough (e.g. borosilicate) variety should be purchased. Cracked and chipped glass equipment should be discarded. Hypodermic needles should not be used (see Section 6.8.3), nor should scalpels and knives except on histological specimens that have been fixed.

6.12.3.4 *Automated equipment* This should be of the 'enclosed system' type. Probes should be shielded to avoid splashing and the effluent should be trapped in closed bottles or discharged at least 25 cm into the waste plumbing system.

There seems to be no need to dedicate to work with hepatitis B and AIDS materials equipment that can be disinfected by passing hypochlorite or glutaraldehyde through it. As some equipment may be damaged in this way the manufacturer should be consulted. It may be necessary to rely on washing through with plain water.

6.12.3.5 *Centrifuges* Sealed buckets should be used.

6.12.3.6 *'Batching' of blood specimens* It is generally accepted that specimens identified as Hepatitis Risk and AIDS Risk should be assembled and tested in automated equipment at the end of a session.

6.12.3.7 *Blood films* As these are 'open' they should be handled with care, using forceps (viruses do not jump!).

Treatment with 70% ethanol or 70% isopropanol is said to kill both viruses, but no information seems to be available about the action of methanol.

6.12.3.8 *Tissues*   All material for histological examination should be fixed before any work is done on it. Small specimens, e.g. needle biopsies, are fixed by formalin within an hour or two but larger specimens may take several hours or days.

Frozen section work on fresh material should be avoided, but if it is essential the cryostat should be well shielded and the operator should wear face protection. After such use the instrument should be raised to room temperature for disinfection with formaldehyde (aldehydes are not effective at low temperatures).

6.12.3.9 *Disinfection*   Hypochlorites and aldehydes are effective against both viruses. Three strengths of hypochlorites are recommended. Dilutions of commercial hypochlorites in water should contain:

- For general use, 1000 ppm available chlorine
- For pipette jars, 2500 ppm available chlorine
- For blood spillage, 10 000 ppm available chlorine.

Hypochlorite solutions should be freshly prepared each day.

Glutaraldehyde (2% solution), is used as purchased. It is preferable to hypochlorite for disinfecting surfaces and metal equipment.

Further information about disinfectants is given in Section 6.14 below.

6.13 MICROBIOLOGICAL SAFETY CABINETS .

Microbiological safety cabinets are intended to protect the worker from airborne infection, which might result from the inhalation of infectious particles released into the air in the course of the work. They do not protect against infections associated with ingestion, injection, skin contact etc. There are three classes, I, II and III.

6.13.1 Class I microbiological safety cabinets

Class I cabinets (to British Standard 5726: BSI 1976 as amended) are specified by the ACDP (1984). Air is extracted from the room, passes over the working area, where it entrains airborne particles, thence through a HEPA filter which removes most if not all of those particles and is then exhausted to atmosphere (Fig. 6.1a).

The effluent air from these cabinets should not be recirculated into the room. When this is 'impossible' however, as happens in some very badly designed laboratories, the ACDP (1984) makes a special provision for the air to be passed through a second HEPA filter and recirculated into the room. This course should be avoided for a variety of reasons, not the least of which is the problem of decontamination (see Section 6.13.6 and Collins, 1988).

There should be a Class I cabinet in each Containment Level 3 laboratory room and one in every diagnostic laboratory suite for the general use of all departments as and when required. These cabinets should be installed in accordance with the recommendations in Appendix D of BS 5726 (BSI, 1976), Clark (1983) or Collins (1988) or they may not protect the worker. The main requirements are that they should not be near to doors and windows or where people walk past them when they are in use, and no draughts or air currents should pass across their working faces. Poor siting and certain air movements may allow infectious particles to return to the room through the open front.

6.13.2 Class II microbiological safety cabinets

Class II cabinets operate on a different principle (Fig. 6.1b). Incoming air does not pass directly over the working area. It is filtered and then some (60–90% according to type) is recirculated down through the cabinet. The working area is therefore 'clean', unlike that in most Class I cabinets. Class II cabinets are officially frowned upon in the UK but are in general use elsewhere. They require more stringent maintenance than Class

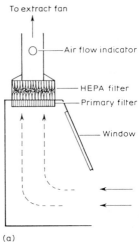

To extract fan

Air flow indicator

HEPA filter

Primary filter

Window

(a)

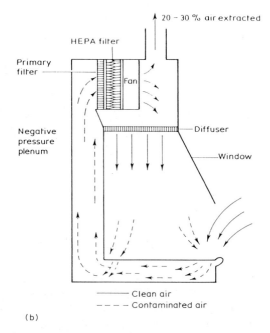

20 – 30 % air extracted

HEPA filter

Primary filter

Fan

Negative pressure plenum

Diffuser

Window

——— Clean air
– – – – Contaminated air

(b)

FIGURE 6.1 Microbiological safety cabinets. (a) Class I; (b) Class II. Reproduced from: Collins, C.H. (1988) *Laboratory Acquired Infections*, by permission of the publishers, Butterworths, London.

I cabinets. For more information about Class II cabinets see Collins (1988) and Clark (1983).

6.13.3 Class III microbiological safety cabinets

Class III cabinets are used in Maximum Containment (Level 4) laboratories and are not considered here.

6.13.4 Working in safety cabinets

All materials and equipment should be placed in the cabinet before commencement of work. It is unsafe to 'put and take' as each time the arms are withdrawn some air, possibly containing aerosols, may be brought into the room. Large items of equipment should be avoided. They compromise the air flow and therefore the safety of the worker.

Work should be done in the rear two-thirds of the cabinet, not near to the front. Work should not commence until the air flow indicator is in the 'safe' position. When the work is completed, arms and equipment should not be withdrawn for about one minute to allow any aerosols to be cleared from the cabinet.

The front closure should be in place whenever the cabinet is not in use.

6.13.5 Testing cabinets

Cabinets should be tested by the manufacturer or a professional organization at regular intervals (e.g. six-monthly) on a contract basis. Tests will include air flow, ability to contain particles and integrity of the filters and will be in accordance with the British Standard (BSI, 1976).

If at any time the air flow indicator falls below the 'safe' position, or local tests with an anemometer give readings of less than 0.75 m/sec, the cabinet should be tested professionally.

6.13.6 Disinfection of cabinets

The working surfaces of safety cabinets should be disinfected after use by wiping them down with glutaraldehyde. Before changing the filters, testing by the manufacturer or contractor, and at regular intervals depending on use (e.g. monthly to three-monthly), the cabinet should be disinfected with formaldehyde (Collins, 1988).

The appropriate amount of commercial formalin solution (specified by the manufacturer according to the cubic capacity of the cabinet) and an equal volume of water should be boiled away on an electric heater (purpose-made models are provided by manufacturers) with the front closure in place and the fan off. When the formalin has boiled away the heater is switched off, the front closure 'cracked' open slightly and the fan switched on. After 15–30 minutes the front closure may be removed and the fan left running for another 30 minutes to clear remaining formaldehyde.

It is very difficult to disinfect the filters of safety cabinets that recirculate air to the laboratory unless the whole room is sealed and left out of action for at least 24 hours (Collins, 1988).

6.14 DISINFECTION AND DECONTAMINATION

'It is a cardinal rule that no infected material shall leave the laboratory' (Collins *et al.*, 1974).

Failure to observe this rule has exposed members of the public to infection and brought unwelcome publicity to several laboratories.

6.14.1 Chemical disinfection

Chemical disinfectants are used in bench discard and pipette jars, for treating surfaces that might be contaminated, and for disinfecting equipment.

6.14.1.1 *Discard jars*   Clear phenolics are used mostly in bacteriology. There are many varieties but only well-known commercial products should be used. They should be diluted according to the manufacturer's directions (usually 2–5% in water). Hypochlorites are usually preferred by virologists, haematologists and clinical chemists. Solutions in water should contain at least 1000 ppm available chlorine (see Section 6.12.3.9).

Discard jars should be refilled daily with freshly-prepared disinfectant dilutions. Older solutions are usually ineffective. Jars should not be overloaded or the efficiency of the disinfectant will be reduced. At the end of each day the jars should be collected. After standing overnight the contents should be emptied through a metal sieve into a bucket for autoclaving or incineration. The jars should be washed in very hot water before re-use.

Methods for testing the efficiency of disinfectants in jars during use are given by Collins (1988) and Collins and Lyne (1984). Results may be very revealing!

6.14.1.2 *Pipette jars* Contaminated glass pipettes cannot usually be autoclaved. They should be left overnight completely submerged in hypochlorite containing 2500 ppm available chlorine and then washed in hot water.

6.14.1.3 *Surfaces and equipment* Glutaraldehyde is the disinfectant of choice. Phenolics leave sticky residues, hypochlorites attack metals and should not be used on moving parts or equipment liable to stress (e.g. centrifuge tubes and rotors). Formaldehyde is used for space disinfection, e.g in safety cabinets (see Section 6.13.6).

6.14.2 Autoclaving

Only those autoclaves that are designed for laboratory use will give satisfactory results. Bottled fluid sterilizers, dressing sterilizers etc. are not satisfactory. Autoclaves that fail and need frequent servicing cause serious problems. Two small autoclaves are better than one monster. Autoclaves should be fitted with flexible internal thermocouple probes so that the temperature *in the load* can be recorded. That in the drain is irrelevant

(Collins, 1988). Even so, occasional tests for effective disinfection, using spore papers or chemical indicators are useful.

Infected waste should be autoclaved at 121°C for 20 minutes. The starting time is when the thermocouple meter reads 121°C, because different autoclaves and different loads take varying times to reach that temperature. (see Collins, 1988 and Gardner and Peel, 1986).

6.14.2.1 *Autoclave containers* Containers must not be too deep or steam will not penetrate the load. They should have solid bottoms so that they do not leak. They may be of metal or plastic. Re-usable articles should be placed directly in these containers but disposable waste is best placed in commercial autoclavable plastic bags. The bags should always be supported in other containers in case they burst under the weight of their contents. They should be left open during autoclaving.

6.14.3 Incineration

Only those incinerators that have efficient after-burners are suitable for the destruction of infectious waste. In others, micro-organisms may be carried up the flue by the draught long before the flames reach them.

Direct incineration of waste, i.e. instead of autoclaving, is satisfactory only if the incinerator is under direct laboratory control. There have been regrettable incidents in which bags of infectious material have failed to reach an incinerator.

6.15 EMERGENCY PROCEDURES

The most likely emergencies are breakages and spillages of cultures. Although these may release aerosols the amount is not likely to be as great as, for example, a centrifuge accident. Readily accessible emergency kits, should be provided and should contain the following:

1. Bottles of different kinds of undiluted disinfectants and empty bottles for preparing dilutions.

2. Paper towels and pieces of strong cardboard about 25 x 10 cm.
3. Forceps for picking up broken glass.
4. An autoclavable dustpan.
5. Boxes or bags to receive glass and other waste.
6. A pack containing disposable gowns, overshoes, face masks, head covers and goggles.
7. Rubber shoes (Wellingtons).

6.15.1 Simple breakage and spillage

Simple ('non-violent') accidents may usually be dealt with by covering the debris, spillage etc. with paper towels and pouring over them a liberal amount of disinfectant. After about an hour, the debris should be cleared up into a box using the dustpan and a strip of card. This is taken away for autoclaving or incineration. A final wash with disinfectant is then all that is required.

6.15.2 Major breakages; release of aerosols

Such accidents include dropping several cultures some distance so that the debris is widely spread, and breakage of tubes in a centrifuge. Large amounts of aerosols may be formed.

The occupants of the room should leave it immediately. The nearest person should switch off the centrifuge. After 30 minutes, during which time most of the larger aerosol particles will have settled, the room may be entered by a senior member of the staff (e.g. the safety officer) wearing appropriate protective clothing, including a respirator if dangerous pathogens have been involved. He will then clear up the mess as above and decide if the room should be fumigated. This is usually done with formaldehyde and requires consultation with the Control of Infection Officer or other professionals.

If a tube has broken in a centrifuge, glutaraldehyde should be used and the debris removed with forceps. The rotor and buckets should be removed for autoclaving and the bowl swabbed with more glutaraldehyde.

6.16 REFERENCES

ACDP (1984) *Categorisation of pathogens according to hazard and categories of containmnent.* Advisory Committee on Dangerous Pathogens. London: HMSO.

ACDP (1986) *LAV/HTLV-III - the causative agent of AIDS and related conditions - Revised Guidelines.* Advisory Committee on Dangerous Pathogens. London: HMSO.

Allen, B.W. (1981) Survival of tubercle bacilli in heat-fixed smears. *Journal of Clinical Pathology* **34**, 719–722.

Allen B.W. & Darrell, A (1983) Contamination of specimen container surfaces during sputum collection. *Journal of Clinical Pathology* **36**, 479–481.

Blaser, R. & Feldman, R.A. (1980) Acquisition of typhoid fever from proficiency testing specimens. *New England Journal of Medicine* **303**, 1481.

BSI (1972) BS 4851 *Medical specimen containers for haematology and biochemistry.* London: British Standards Institution.

BSI (1975) BS 5213 *Medical specimen containers for microbiology.* London: British Standards Institution.

BSI (1976) BS 5726 *Specification for microbiological safety cabinets.* London: British Standards Institution.

CDC (1974) *Lab Safety at the Centers for Disease Control.* Atlanta: CDC.

CDC (1985) Revision of case definition of acquired immunodeficiency syndrome for national reporting - United States. *Morbidity and Mortality Weekly Reports* **34**, 575–578.

CDC (1986) Human T-cell lymphotrophic virus type III/lymphadenopathy-associated virus: agent summary statement. *Morbidity and Mortality Weekly Reports* **35**, 540–549.

CDC/NIH (1984) *Biosafety in Microbiological and Biomedical Laboratories.* Washington: Government Printing Office.

Clark, R.P. (1983) *The Performance, Installation, Testing and Limitations of Microbiological Safety Cabinets.* London: Science Reviews. pp. 67–72.

Collins, C.H. (1988) *Laboratory Acquired Infections.* 2nd edn. London: Butterworths.

Collins, C.H. (1984) In perspective: laboratory-associated infections. *Abstracts of Hygiene and Communicable Diseases* **59**, R1–R15.

Collins, C.H. & Kennedy, D.K. (1987) Microbiological hazards of needlestick and sharps injuries: a review. *Journal of Applied Bacteriology* **62**, 385–402

Collins, C.H. & Lyne, P.M. (1984) *Microbiological Methods.* London: Butterworths. 5th edn. p.29.

Collins, C.H. Grange, J.M. & Yates, M.D. (1986) *Organization and Practice of Tuberculosis Bacteriology.* London: Butterworths.

Collins, C.H., Hartley, E.G. & Pilsworth, R. (1974) *The Prevention of Laboratory Acquired Infections.* Public Health Laboratory Service Monograph No. 6. London: HMSO.

Darlow, H.M. (1969) Safety in the microbiological laboratory. In *Methods in Microbiology 1.* Eds. J.R. Norris & D.N. Ribbons. London: Academic Press. pp. 169–204.

Darlow, H.M. (1972) Safety in the microbiological laboratory; an introduction. In *Safety in Microbiology.* (Eds. D.A. Shapton & R.G. Board. London: Academic Press. pp. 1–19.

DHSS (1978) *A Code of Practice for the Prevention of Infection in Clinical Laboratories and Post-mortem Rooms.* London: HMSO.

Gardner, J.F. & Peel, M. (1986) *Introduction to Sterilization and Disinfection.* London: Churchill-Livingstone. pp. 64-82.

Ghosh, H.K. (1982) Laboratory-acquired shigellosis. *British Medical Journal* **285**, 695–696.

Grist, N.R. (1981) Hepatitis and other infections in clinical laboratory staff. *Journal of Clinical Pathology* **34**, 655–658.

Grist, N.R. & Emslie, J.A.N. (1987) Infections in British clinical laboratories 1982–1983. *Journal of Clinical Pathology* **40**, 826–829.

Hospital Infection Society (1985) Report of a Working Party. Acquired immune deficiency syndrome: recommendations. *Journal of Hospital Infection 6* (Supplement C), 67–80.

HSAC (1985) *Safety in health service laboratories: hepatitis B.* Health Services Advisory Committee. London: HMSO.

HSAC (1986) *Safety in health service laboratories: the labelling, transport and reception of specimens.* Health Services Advisory Committee. London: HMSO.

Kurl, D.N. (1981) Laboratory-acquired human infection with Group A type 50 streptococci. *Lancet* **ii**, 752.

Levy, J.A., Kaminsky, L.S., Marrow, J.W. *et al.* (1985) Infection by the retrovirus associated with the acquired immune deficiency syndrome. *Annals of Internal Medicines* **103** 694–699.

NIH (1974) *National Institutes of Health Biosafety Guide.* Washington: NIH.

Phillips, G.B. (1961) Microbiological Safety in US and Foreign Laboratories. Report No. 35, US Army Medical Corp. Washington: US Army.

WHO (1979) Safety measures in microbiology. Minimum standards of laboratory safety. *World Health Organisation Weekly Epidemiological Record* No. 44: 340–342.

WHO (1983) *Laboratory Biosafety Manual.* Geneva: World Health Organisation.

7 HEALTH CARE IN
THE LABORATORY

A.E. Wright

In occupational health, as in other aspects of our daily life, public opinion often foreshadows legislation. Thus concern at the hazards of certain occupations in the UK found expression in the report of the Robens Committee (Report, 1970-72) and led to the *Health and Safety at Work etc. Act* of 1974. This Act was designed to provide comprehensive cover for persons at work and closed many of the loopholes that had persisted from earlier legislation. In particular, it covers educational establishments and research and other laboratories and places a responsibility on an employer to 'ensure, as far as is reasonably practicable the health, safety and welfare at work of all his employees'. The Act is administered by the Health and Safety Commission and enforced by the Health and Safety Executive through its inspectors. It follows from this that those in managerial charge of laboratories have a responsibility not only to ensure safe practices in their day-to-day work but also to have a direct concern for the health of their staff.

It should not be assumed that the need for legislation implied that the health of those who worked in health service establishments had previously been ignored. This is far from the case. Interest in safety in laboratories has been appreciable over the years and is discussed in Chapter 1. Occupational health care in the form of preventive medicine has had a less than lustrious history in the National Health Service (NHS). Pressure from individuals and a number of professional organizations resulted in a Report (1968) - The Tunbridge Report - which was entitled *The Care of the Health of Hospital Staff*. At that time the NHS had been in existence for 20 years without any real steps being taken to create an occupational health service. Some hospital authorities had created such a service but these were the exceptions. Today such provision remains patchy, and ranges from the specialist physician with supporting staff to a part-time appointment, usually of a general practitioner, for a varying number of sessions each week. Less than ten years ago the HSE (1978), in a pilot study on working conditions in the medical service, reported that '...some resistance to an occupational health department concerned with prevention rather than just treatment.' It is to be hoped that this attitude is changing, but whatever the position in the NHS in the UK, the head of a clinical laboratory has a clear moral and legal responsibility to care for the health of the staff. The Act also lays a responsibility upon the individual to look after their own health and safety and that of fellow workers.

7.1 MEDICAL FITNESS FOR WORK

Regulations concerning fitness to work vary with the employer: some require a full medical examination and others merely the completion of a questionnaire. The *Code of Practice for the Prevention of Infection in Clinical Laboratories and Post-mortem Rooms* (DHSS, 1978) - known as the 'Howie Code' - requires a prospective laboratory employee to have a medical examination or to provide a statement of fitness for the job from a general practitioner. Not all are enamoured of routine medical examinations and in a large organization the necessary arrangements can lead to considerable delays in appointments, due to staff turnover. Often, a carefully compiled medical questionnaire is adequate to pick out those who require further investigation. A form carrying a warning that false statements will lead to dismissal is sufficient to ensure that these rarely appear.

Whichever course is taken, attention should be paid to height and weight in relation to sex and age. Differences approaching 10% should lead to further investigation. It is wise to explore histories of allergies and skin diseases as high prevalence rates of allergy to small animals have often been reported (Agrup *et al.* 1986). One advantage of a medical examination is the way in which a candidate responds to questions, as this will give the medical examiner an opportunity to make personality judgements. This may be important in laboratory work, as the introspective individual may be unable to overcome his apprehension at having to handle material from patients suffering from, e.g. AIDS, or hepatitis. Such an attitude may be detected and the candidate not employed. It is obviously important to judge the physical capabilities of the candidate if the post involves heavy work that may be required in a large animal house. Questions should be asked about medication, and those who are immunosuppressed should not, of course, be employed in clinical and biomedical laboratories. Evidence of other forms of drug taking may not be so easy to discover but an attempt should be made.

Questions should also be asked about vaccination and the candidate's attitude to such protection. The fact that a candidate has no record of vaccination may more truly reflect the attitude of the parents than of a young person applying for a first appointment. If BCG has been given a scar will be visible. The Howie Code makes a large chest radiograph mandatory unless one has been taken during the previous 12 months.

Minor disabilities do not necessarily debar a good candidate; enthusiasm and a good academic record may be more important. It is often possible to employ a registered disabled person in some capacity in a large laboratory and indeed this is a requirement of the *Disabled Persons Employment Acts*. These place an obligation on certain employers for 3% of their workforce to be recruited from among registered disabled persons.

The nature of the work should be explained to the candidate, either at the interview or during the medical examination. Although it will not be possible to explain in any detail the nature of the hazards of working in a laboratory some indication should be given in order to test the reaction of the candidate. If not already explained the need for vaccination is a natural corollary to this discussion, as is the information that he or she may be required to work with animals.

Some minor disabilities might be better termed variations from normal and considered as such. Colour blindness for example, may be present in about 8% of men and at a very much lower level in women. This does seem to hamper laboratory workers who recognize objects under the microscope as much by their structure as by their colour. Standard tests, depending for results on colour changes, also seem to be 'sensed', partly by comparison with positive and negative controls. Many epileptics can also be successfully employed in laboratories. Modern drugs are now available to keep many such sufferers under control for long periods. In addition those who do suffer a rare attack, brought on by some stress, will have adequate premonitory signs and will thus be able to avoid a disaster.

7.2 INDUCTION

Small laboratories cannot allow themselves the luxury of time to introduce a new employee to work in a formal manner. This is more usually done by allocating the newcomer to a senior member of the staff who is known to have a balanced view of life in general and of the safety aspects of the laboratory work in particular. In larger laboratories time, often several days, may be allocated for an induction course and during this time a number of points should be emphasized:

1. A medical record card (or computer record) should be completed, designed to display the employee's immunization history. The computer system has the advantage that a print-out of those requiring booster doses can be available on a monthly basis. This card should also carry the names of the next-of-kin and the employee's general practitioner.
2. A letter should be sent to the employee's medical practitioner indicating that the individual has commenced work in a clinical or biomedical laboratory and that he will shortly be offered immunization against a number of infectious agents unless the practitioner prefers to do these himself.
3. The employee should be issued with a personal record card showing his name and address, the name and address of his doctor and those of the senior medical staff of the laboratory.

The card is designed to be shown to the attending physician if the employee is taken ill, thus ensuring that occupational hazards are considered in the differential diagnosis.

4. By agreement with the employee, a sample of blood should be collected to be screened for antibodies to rubella (in females), hepatitis B and HIV. This specimen should be stored for retrieval at a later date if required.

5. The employee should be conducted round the laboratory and instructed in the safety precautions that are in operation. The legal aspects of safety should be explained and the workings of such apparatus as microbiological safety cabinets made clear.

7.3 VACCINATION PROGRAMMES - PRACTICE AND PROBLEMS

If the employment medical examination of new staff has been properly carried out there should be no contraindications to vaccinations. More established staff may, over the years, develop contraindications and each case must be considered on its merits. Similarly, the interview will have eliminated those who are not prepared to accept vaccination, although problems do occasionally present themselves in this area. For instance, the Howie Code implies that immunization procedures should be made a 'condition of employment in a laboratory or part of a laboratory.' It goes on to say, however, that this applies only 'if the nature of the work and the risks to employees and their contacts merit such action'. This seems to allow the laboratory head a degree of latitude depending on the infectious agents being handled, although such latitude has never been tested in the courts in the UK. The interpretation of such advice is discussed under various headings below. The suggestion in the Howie Code that 'Staff already in post who refuse the appropriate immunization must accept transfer to other work in order to remove potential risks to their colleagues and their families' is rarely a practical solution.

The following vaccinations may be indicated:

7.3.1 Tuberculosis

Anyone who has had a BCG vaccination while at school will probably show a scar and this may be taken as sufficient evidence of protection against infection with *Mycobacterium tuberculosis*. If no such scar is seen skin testing is indicated and if this is negative, BCG administered. Tuberculin positivity develops in from 3 to 12 weeks so newly vaccinated subjects should not be permitted to work with tuberculous material during this period. Some care should be taken if the employee was previously vaccinated in India as a recent study in south India (WHO, 1979) showed no protective effect in over a quarter of a million people in a 7½ year period. There may have been a number of reasons for this: the efficiency of the vaccine or an impaired response in a poorly nourished infant (Lyon, 1986). Hence revaccination may be indicated. Should there be any doubt about vaccination status the employee should have a skin test as BCG is not by any means administered universally, even in Western Europe. Conversely those who receive BCG unnecessarily or inexpertly may be left with a discharging ulcer which takes many weeks to heal. There is good evidence that in the UK BCG gives long-lasting immunity even if skin sensitivity wanes.

It has been shown by several workers that the incidence of tuberculosis in laboratory staff in the UK was higher than that in the general population but no recent studies have been done. The most recent publication (Harrington and Shannon, 1976) is now out of date and as the incidence of the disease in the general population has fallen it will no doubt have fallen in laboratory workers. The incidence in the UK makes it extremely difficult, on statistical grounds, to obtain accurate figures for those who work in clinical laboratories. In addition, as the overall incidence falls, so it becomes more difficult to obtain meaningful figures for those exposed to different degrees of hazard - for example those working with mycobacteria in a reference laboratory compared with those in, say, a clinical chemistry laboratory or a post-mortem room. This fall in the incidence of the disease may lead to the abandonment of BCG vaccination in school children in the UK. The history of tuberculosis in laboratory workers however, is such that they are likely to be seen as a special case and are likely to continue to receive the vaccine (Collins, 1982).

7.3.2 Enteric fever

Paratyphoid A and B are now rare in the UK and only about 200 cases of typhoid are diagnosed each year. As only one or two of these become chronic excretors the average clinical laboratory encounters such infections infrequently. In the USA however, there have been a number of cases of laboratory-acquired typhoid fever, associated with proficiency testing specimens (MMWR, 1979).

It is still taught that workers in clinical laboratories should be vaccinated every three years. Fortunately the TAB vaccine, which gave rise to many side effects, is no longer used. It has been superseded by the monovalent typhoid vaccine. This gives fewer side effects, either because of the lower dose of organisms or the absence of the paratyphoid element, and is much more acceptable. At one time those who had been vaccinated with TAB at frequent intervals complained of more severe reactions to the vaccine as they grew into middle age and it became the custom not to insist on booster doses for anyone who had received frequent injections. This, coupled with the fact that the vaccine is probably only 70–90% effective means that there is a certain sense of false security about the protection of staff against typhoid in the UK. In view of this it may be unreasonable to insist on the vaccination of office staff and other workers in smaller laboratories who do not directly handle potentially infectious material. Clearly there is room here for judgements to be made but these should not be extended to the staff of reference laboratories which receive many strains for identification and typing, nor to enteric research laboratories.

Further progress in the use of the oral typhoid vaccine using the Ty2/s strain is awaited with interest, but at present the vaccine is given subcutaneously (s.c.) or intramuscularly (i.m.) in doses of 0.5 ml. As an alternative 0.1 ml is given intradermally (i.d.), a method that gives fewer side effects. A second dose is required at four to six weeks. Booster doses are given every three years.

7.3.3 Tetanus

Vaccination against tetanus should be encouraged for all the population and especially for those who engage in physical sports. It is not a disease of laboratory workers although cases have occurred in specialized laboratories. Vaccination should be offered to all NHS staff and compliance encouraged. Many people in the UK will already have received tetanus toxoid in the triple vaccine given in childhood so all that will be required is a booster dose. The full course is three injections of 0.5 ml of the adsorbed tetanus toxoid given i.m. or deep s.c. at intervals of one to eight weeks followed by a third dose at four to six months. A booster dose should be given five to fifteen years later.

7.3.4 Hepatitis B

Hepatitis is now known to be caused by several different viruses and although all are infectious the mode of transmission of infection differs, as does the natural history of the disease. Serum hepatitis, now called hepatitis B, is the most feared by laboratory workers, although the incidence in the UK is much lower than in many other parts of the world. It is the HBsAg positive patients who also possess the 'e' antigen without any corresponding antibody whose blood and secretions may present a hazard to the laboratory worker. Precautions are described in Section 6.12.

If accidents do occur hepatitis B immunoglobulin can be administered if the recipient is seronegative and the specimen is from a known seropositive individual.

Those who are in continual contact with material from hepatitis B positive patients or who handle material from other known high risk patients such as drug abusers should be offered hepatitis B vaccine. This consists of surface antigen particles that have been carefully purified and processed to inactivate the virus. It has been used successfully for large groups of people throughout the world and there is no doubt about its safety. What is a problem is that three doses of 1.0 ml i.m. are required initially and at one and six months, and the cost at the time of writing is of the order of £65.00. A further problem is that although the response to vaccination may exceed 90% this is not adequate for the group of workers under consideration. The antibody response must be checked after the initial course of vaccine and frequently (every two years) thereafter, as antibody may not persist. Such a procedure is fraught with difficulties in the UK and may be insurmountable in developing countries. A genetically engineered hepatitis B vaccine is at present being assessed and it may prove to be more effective*.

Not all laboratory workers are prepared to accept hepatitis B vaccination and this can pose a difficult problem to the head of the laboratory. If vaccination is refused it is not unreasonable to request the individual to sign a statement indicating that refusal. In the case of hepatitis it is probably not reasonable to make this a condition of employment, for as indicated elsewhere in this book, a competent individual should be able to do this kind of work without becoming infected. In most laboratories it would not be administratively possible to allow those who refuse vaccination to work elsewhere. This situation has not been tested in the courts in the UK but as the incidence of infection is low, a competent worker who prefers not to be vaccinated should be allowed to work with hepatitis B material providing the hazards are clearly understood by him and the workload is modest.

7.3.5 Rubella

All women of child-bearing age should be vaccinated against rubella and in addition, laboratory staff should have their sera tested for antibodies to ensure that they are adequately protected. No reliance should be placed on a history of an illness resembling rubella as a number of virus infections can give rise to clinical illness resembling this disease.

---

* This vaccine, a yeast-derived recombinant DNA vaccine is now available. The indications for use and administration remain the same but the cost of both vaccines is now halved.

The uptake of this vaccine is inadequate in the UK and criticism is becoming apparent at the whole concept of vaccinating females but not males, thus allowing wild virus to continue to circulate in the population. No opportunity to vaccinate people who present themselves at occupational health departments should therefore be ignored.

The vaccine is given in one dose of 0.5 ml s.c. Contraceptive measures should be taken for 12 weeks after vaccination.

7.3.6 Poliomyelitis

Vaccination against poliomyelitis is relatively safe and effective and all laboratory workers should be offered such protection. Although the disease is now rare in the UK it is not uncommon in other parts of the world. Specimens of faeces from abroad or from recent arrivals in this country may contain the virus and be handled in UK laboratories. Infection is by ingestion and therefore should not occur (see Section 6.8.2) but it is as well that all should be protected.

The oral attenuated vaccine multiplies in the gut providing local gut immunity. Three doses are required at intervals of six to eight weeks and four to six months. Contraindications include pregnancy, gastrointestinal disturbances and hypersensitivity to any of the antibiotics used in the preparation of the vaccine. It is generally wise to vaccinate whole families at a time if there are young children, for there is the possibility that the virus may increase in neurovirulence by passage through the gut. Such contact vaccine-associated poliomyelitis is, however, very rare.

Inactivated poliomyelitis vaccine, given in a dose of 0.5 ml by deep s.c. or i.m. injection, is also available.

7.3.7 Cytomegalovirus

Like rubella, cytomegalovirus may cause intrauterine infections ranging from severe malformation causing abortion, to microcephaly and mental defects, the evidence for which may be delayed. Inapparent infection may be common, resulting in immunity in some adults. We have no evidence that this disease is a problem in laboratory workers but reference laboratories must obviously keep the effects of such infection in mind when training staff. Vaccination has been used.

7.3.8 Toxoplasma

Toxoplasma gondii presents problems similar to those with cytomegalovirus but no vaccine has yet been developed.

7.3.9 Diphtheria

Diphtheria is now a rare disease in the UK because of the highly successful vaccination programme applied during early childhood. It is rare for adults to receive booster doses, as formol toxoid should be given to adults only after they have been tested for susceptibility (Schick test). This procedure can now be avoided by using the (adsorbed) diphtheria vaccine for adults (Swiss Serum and Vaccine Institute, Berne, distributed by Regent Laboratories, London). The dose is 0.5 ml by i.m. or deep s.c. injection.

7.3.10 Rabies

Rabies is unlikely to be a hazard in any other than research laboratories. Vaccines are available and should be offered to those working with the agent. Until recently antibody response to rabies vaccine was poor but the human diploid cell vaccine (Merieux Institute) has proved itself in the field. The dose is 1.0 ml by deep s.c. injection initially, at one month and at six to twelve months. Booster doses may be required at intervals of three years. The vaccine should not be given intradermally unless arrangements can be made to establish the titre of the antibody produced. A new tissue culture rabies vaccine using vero-cells is currently receiving field trials and may be marketed more cheaply than the present vaccine.

7.3.11 Anthrax

There have been a number of cases of laboratory acquired anthrax (see Collins, 1988), and some deaths have been due to the inhalation of aerosols produced during work on bacteriological warfare. The vaccine is prepared from the Sterne strain of *Bacillus anthracis* and is suitably inactivated and alum precipitated. Four doses of 0.5 ml are given by deep s.c. or i.m. injection, the first three at intervals of three weeks and the last one six months later. There may be some local discomfort and swelling and occasionally lymphadenopathy and pyrexia. The vaccine is prepared by the Public Health Laboratory Service for the Department of Health and Social Security.

7.3.12 Botulism

Reports of laboratory infections with *Clostridium botulinum* are rare (Holyer, 1962) but as the toxin may be absorbed from the nasal mucosa (Petty, 1965) those who work with the agent should be protected. The vaccine, which appears to produce good antitoxin levels, is a pentavalent toxoid aluminium phosphate-adsorbed vaccine containing toxoids from formalin-inactivated types A,B,C,D and E. The dose is 0.5 ml by deep s.c. injection at intervals of two and twelve weeks with a booster dose at twelve months. It may be obtained from the Centers for Disease Control, Atlanta GA, USA and is available only under strict conditions.

7.3.13 AIDS

There is no vaccine at present. For precautions see Section 6.12.

7.4 ABSENCE DUE TO ILLNESS

Members of the staff must inform the head of the laboratory when they are absent because of illness. This should be done by a relative or friend on the first morning of their absence. If no message is received then arrangements should be made to visit the home of the employee to ensure that illness is not due to a laboratory infection.

7.5 PREGNANCY

Although all causes of abortion and malformation of babies are not understood we have clear evidence concerning the role of rubella and some evidence concerning that of cytomegalovirus and mumps. Women hoping to become pregnant and those in the early weeks of pregnancy should not work in virus laboratories. A change of occupation can usually be arranged. Although such a move in a small clinical laboratory may cause inconvenience there is little doubt that any untoward occurrence would be blamed on the laboratory. Such an incident would be difficult to refute.

7.6 CHEST X-RAYS

All staff who handle known or suspected tuberculous material should have an annual large film chest X-ray and other staff should be X-rayed every three years.

7.7 WORKING WITH ANIMALS

Isolation methods used by microbiologists are now so much improved in sensitivity that animals are less commonly used in diagnostic work. Some minimum facilities in selected laboratories have to be maintained but most large animal houses are now found only in research establishments.

As in the laboratory, safe procedures should be laid down and observed. Animal houses should have adequate changing and showering facilities as well as a staff room for rest periods. The staff should be made aware, not only of the hazards presented by the organisms being investigated, but by the diseases classed as zoonoses. There are also additional hazards associated with the manipulation of animals as the operator may be using sharp instruments. The skin of an animal should be always disinfected after inoculation (as well as before) to prevent the creation of aerosols (Hanel and Alg, 1955).

Inoculated animals may also be sources of infection from urine, faeces and droplets from coughing and sneezing. Animal houses are often warm and at times may be dusty so allergic individuals should not work in them. Vaccinations are important, especially against tuberculosis and tetanus and in specialized laboratories the animal house staff should have protective inoculations to cover the diseases being investigated.

7.8 CONCLUSIONS

In many laboratories in district general hospitals in the UK the care of the health of laboratory workers is in the hands of the hospital occupational health department. From the point of view of the administration this is a very tidy solution and in many instances it will function well. There are very definite hazards however, associated with laboratory work and work with animals and if the health of these workers is to be properly supervised the medically qualified pathologist, and especially the microbiologist has a part to play.

7.9 REFERENCES

Agrup, G., Belin, L., Sjosted, L. & Skerfring, S. (1986) Allergy to laboratory animals in laboratory technicians and animal keepers. *British Journal of Industrial Medicine* **43**, 192–198.

Collins, C.H. (1982) Laboratory-acquired tuberculosis. *Tubercle* **63**, 151–155.

Collins, C.H. (1988) *Laboratory Acquired Infections.* 2nd edn. London: Butterworths.

DHSS (1978) *A Code of Practice for the Prevention of Infection in Clinical Laboratories and Post-mortem Rooms.* London: Department of Health and Social Services.

Hanel, E. & Alg, R.L. (1955) Biological hazards of common laboratory procedures. II. The hypodermic syringe and needle. *American Journal of Medical Technology* **21**, 343–346.

Harrington, J.M. & Shannon, H.S. (1976) Incidence of tuberculosis, hepatitis, brucellosis and shigellosis in British medical laboratory workers. *British Medical Journal* **1**, 759–762.

Holyer, E. (1962) cited by Petty q.v.

Health and Safety Executive (1978) *Pilot Study: Working Conditions in the Medical Service*. London: Health and Safety Executive.

Lyon, A.L. (1986). Letter. *British Medical Journal* **2**, 1526.

MMWR (1979) Laboratory-associated typhoid fever. *Morbidity and Mortality Weekly Reports* **28**, 44.

Petty, C.S. (1965) Botulism: the disease and the toxin. *American Journal of Medical Sciences* **249**, 345–350.

Report (1968) *The Care of the Health of Hospital Staff*. (The Tunbridge Report). London: Central and Scottish Health Services Council.

Report (1970–1972) *Safety and Health at Work*. Report of a Committee (Chairman Lord Robens). London: HMSO.

WHO (1979) Tuberculosis prevention trial: Trial of BCG vaccines in South India for tuberculosis prevention: first report. *Bulletin of the World Health Organisation* **57**, 819–827.

8 FIRST AID IN THE
LABORATORY

W.J. Gunthorpe

Most of the accidents that occur in laboratories are of a minor nature and simple first aid treatment is usually adequate, provided that the possibility of infection is kept in mind. Serious accidents do occur however, and by no means all clinical and biomedical laboratories are near to hospital casualty departments, or even have doctors who are able to give correct treatment. It is therefore necessary for some members of the staff of such places, preferably those who already have training in first aid, to know what should be done when there is a serious laboratory accident that is outside the scope of the standard first aid manuals. There is also a statutory requirement for employers to ensure that adequate first aid arrangements are provided.

First aid is given to an injured person in the following circumstances:

1. To preserve life, prevent deterioration of the injured person's condition and promote recovery until medical attention is available.
2. To treat minor injuries which do not require further treatment by a medical practitioner or a nurse.

The *Health and Safety (First Aid) Regulations, 1981*, came into operation in the UK in 1982 and includes an *Approved Code of Practice*. The information given in this chapter is in accordance with those regulations but is restricted to the kind of injuries that may arise from laboratory accidents. For details of other kinds of accidents and emergencies the reader is referred to the *First Aid Manual* (St. John's Ambulance, the British Red Cross Society and the St. Andrew's Ambulance Association, 1982) and *First Aid at Work* (Order of St. John, 1982).

8.1 FIRST AID ARRANGEMENTS IN THE LABORATORY

8.1.1 Administrative considerations

Administrative considerations will depend on the number of employees, the nature of the work, in-house medical and nursing facilities and the distance of the place of work from hospitals and ambulance services. Arrangements must be made about accident recording, notification of serious accidents and laboratory-acquired infections to the Health and Safety Executive (HSE) in accordance with the *Reporting of Injuries, Diseases and Dangerous Occurrences Regulations 1985* (RIDDOR).

Regular checks should be made of first aid boxes, eye-wash bottles, stocks of dressings and materials. Suitable rest rooms should be provided. An up-to-date list of qualified first-aiders should be kept. Refresher courses for first-aiders should be held at regular intervals.

Simply-worded notices, indicating the whereabouts of first aid equipment and the names, telephone numbers and normal workplaces of first-aiders should be displayed in prominent places. The telephone numbers of emergency services, ambulances, hospitals etc. should be included.

8.1.2 Choice and numbers of first-aiders

First-aiders should be chosen with care. They need not necessarily have a paramedical or nursing background but should of course, have received training from one of the First Aid Associations (see Section 8.1.3). Ideally such people should be able to remain calm and authoritative during emergencies.

First-aiders should generally be people who remain in one place during most of their working day. Unless paging devices are in use, peripatetic workers should not be chosen because in serious accidents the first three minutes after the accident are vital in the saving of life.

In laboratories there should be one first-aider for every 50–150 employees. There is no statutory requirement for first-aiders in places where there are less than 50 employees.

8.1.3 Training

First-aiders should have been trained and have passed an examination after attending a course of instruction approved by the HSE. Such courses are provided by the St. John's, St. Andrew's and Red Cross organizations. These give all the basic information and practice and train people from all walks of life to do simple things well. Where additional skills are required because of the nature of an employer's business then additional training will be necessary and this is usually arranged by the appropriate occupational health service. First aid treatment for the majority of laboratory accidents however, is contained in the training manuals of the first aid organizations and is supplemented below. Specialist occupational first-aiders are not usually required in clinical and biomedical laboratories.

8.2 FIRST AID FACILITIES

8.2.1 First aid boxes

First aid boxes may be of wood, metal or plastic but must protect the contents from damp and dirt. They should be marked clearly with a white cross on a green background and placed in a prominent and easily accessible place.

The contents, as specified by the Health and Safety Executive under the *First Aid Regulations* should be those listed in Table 8.1 and the box should contain nothing else. The Guidance Card, from the *Approved Code of Practice for First Aid at Work* (HSE 1985) and included in the contents, gives simple details of priority first aid treatment.

TABLE 8.1 Contents of first aid boxes

Soap and water and disposable drying materials, or suitable equivalents, should be available. Where tap water is not available, sterile water or sterile normal saline, in disposable containers each holding at least 300 ml, should be easily accessible, and near to the first aid box, for eye irrigation. Sufficient quantities of each item should always be available in every first aid box or container; at least the numbers of each item shown in the table below should be provided:

Item	Number of employees				
	1–5	6–10	11–50	51–100	101–150
Guidance card	1	1	1	1	1
Individually wrapped sterile adhesive dressings	10	20	40	40	40
Sterile eye pads, with attachment	1	2	4	6	8
Triangular bandages	1	2	4	6	8
Sterile coverings for serious wounds (where applicable)	1	2	4	6	8
Safety pins	6	6	12	12	12
Medium sized sterile unmedicated dressings	3	6	8	10	12
Large sterile unmedicated dressings	1	2	4	6	10
Extra large unmedicated dressings	1	2	4	6	8

From the *Approved Code of Practice for First Aid at Work* (HSE, 1985). Reproduced by permission of the Health and Safety Executive.

8.2.2 Washing facilities

Facilities should include running water, soap and disposable drying materials (e.g. paper towels).

8.2.3 Eye wash facilities

These should be available in all areas where eye injuries might occur. Eye irrigation stations, plumbed into the water services may be provided, but a length of rubber tubing on a cold water tap is usually as effective. If tap water is not available for eye irrigation after a chemical splash, eye-wash bottles containing at least 300 ml of sterile water or sterile physiological saline should be readily available. Refillable bottles are not recommended. If they are not checked regularly bacteria (especially *Pseudomonas aeruginosa*) may grow and use may result in infected eyes. Algae may also grow. Commercial eye wash outfits are available (see Fig. 8.1). The minimum number of eye wash bottles is shown in Table 8.2.

TABLE 8.2 Minimum number of eye wash bottles

	Number of employees			
	1–10	*11–50*	*51–100*	*101–150*
Sterile water or saline in disposable containers (where tap water is not available)	1	3	6	6

From the *Approved Code of Practice for First Aid at Work* (HSE, 1985). Reproduced by permission of the Health and Safety Executive.

8.2.4 Protective clothing

Under normal circumstances protective clothing is not needed for first aid staff but disposable surgical gowns, aprons and gloves may be provided for cases in which there is much bleeding. Such equipment should be placed close to the first aid boxes.

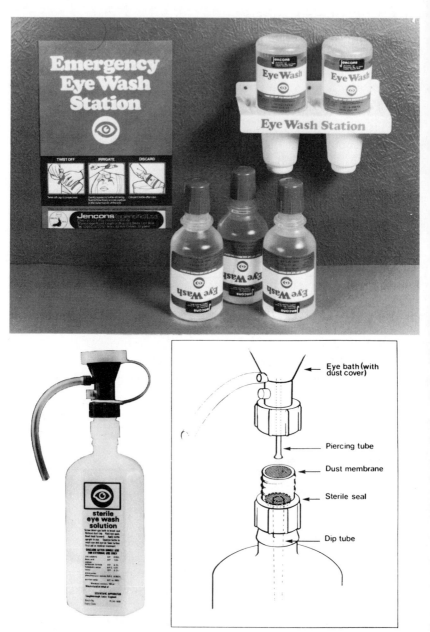

FIGURE 8.1 Eye wash stations (Jencons).

8.2.5 Protective equipment

Brook's airways or similar devices are helpful provided that the first-aider has been trained in their use. The valve type is valuable in cases of asphyxiation caused by poisonous or corrosive chemicals. Respirators are not recommended for the rescue of casualties, e.g. from gas-filled rooms, except by people who have been specifically trained to use them.

8.2.6 First aid rooms

First aid rooms should be sited where they are easily accessible to the personnel of emergency services (ambulances, police etc.) and to both ambulant, wheelchair and stretcher cases. They should also be easily accessible to the vehicles of emergency services. In addition to the first aid materials mentioned above they should also be equipped with:

1. a couch or bed with bedding;
2. chairs;
3. smooth-topped benching;
4. a sink with hot and cold running water;
5. soap, a nailbrush and towels (cloth and disposable);
6. a waste bin;
7. storage cupboards and drawers.

In addition, if the room is used by nursing and/or medical staff, professional equipment will be required which they will specify according to the nature of the casualties that may be expected. The room should be used for no other purpose.

8.3 EMERGENCY FIRST AID

The term emergency first aid is used here to distinguish between certain immediate life-threatening accidents and occurrences, as distinct from simple and less urgent injuries such as minor cuts. A first-aider might not be immediately available and it is therefore incumbent on other laboratory staff to act promptly. Even placing an unconscious casualty in the correct position may save life.

In addition accidents with some chemicals and infectious agents, although not immediately life-threatening, may have serious sequelae if prompt action is not taken. Simple treatment is more likely to be effective than a series of complicated procedures - some of which are likely to be forgotten - or the use of specific antidotes.

8.3.1 Attending a casualty

Before first aid is given the situation should be assessed and the following questions should be asked:

1. Is the casualty still at risk from the cause of the injury, e.g. still holding a piece of electrical equipment or in a poisonous atmosphere? The first-aider should not become another casualty!
2. Can the patient be moved to a safe area without aggravating his injuries or endangering the first-aider?
3. Has help been summoned?
4. What has happened? The casualty may be able to answer that question if he is conscious, or there may be witnesses. There may be evidence of broken glass, spilled chemicals or leaking gas.
5. Are there signs of injury? The patient or witnesses may be able to give this information, or the first-aider may have to examine the patient. It is important to reassure the patient if conscious, to explain that medical assistance is on its way, and to avoid stressing the extent of the injuries.

At the same time treatment should be commenced and this should be in the correct order. A patient suffering for example from blood loss, may become unconscious. The condition may deteriorate further by respiratory and then cardiac failure resulting in death. Unconsciousness must always be treated first and the first-aider must ensure that the patient has a clear airway and can breathe. If he or she is not breathing artificial respiration must be applied. The next priority is the control of bleeding. Wounds should be dressed and the affected parts immobilized. Local causes of injury such as chemical splashes and heat should be neutralized or removed, but glass in the eyes or wounds should be left unless it is easily removed. Pain should be relieved if possible and the patient protected from heat, cold and the pressure of onlookers. The first-aider must remain with the patient until professional assistance arrives and he must be vigilant for signs of deterioration in the injured person's condition.

If there is more than one casualty the first aid treatment should always be in the order of (a) unconsciousness, (b) bleeding (c) other injuries. Help from other people in monitoring unattended victims will enable the first-aider to alter priorities if necessary.

8.4 ASPHYXIA

There may be several causes of asphyxia:

1. Blockage of air supply to the lungs, e.g. by physical obstruction, gas or smoke, compression of the throat or chest, fits.
2. Damage to the central nervous system, e.g. by electric shock, certain poisons.
3. An oxygen-deficient atmosphere.
4. Haemoglobin blockage, e.g. by carbon monoxide, cyanides.

When a person becomes unconscious the tongue loses muscle tone. In the supine position the tongue will fall and obstruct the airway.

The important procedures in preventing death from asphyxia are to maintain a clear airway, keep the patient breathing and his blood circulating.

8.4.1 Treatment

The airway must be cleared of any debris and foreign bodies and restrictions on the neck removed. Extending the head and neck and tilting the head backwards will move the tongue and unblock the trachea. Often this will be sufficient to enable the patient to breathe and he or she can then be placed in the recovery position (Fig. 8.2). If the patient fails to breathe, mouth-to-mouth resuscitation should be given. If asphyxiation was caused by chemicals the first-aider must avoid contact, e.g. with the patient's mouth and alternative methods should be used.

FIGURE 8.2 Recovery position. Reproduced from the *Approved Code of Practice for First Aid at Work* (HSE, 1985) by permission of the Health and Safety Executive.

8.4.2 Mouth-to-mouth resuscitation

The following procedure should be followed (see Fig.8.3):

1. Extend the neck and tilt the head backwards.
2. Support the head by the chin with one hand and pinch the nose.
3. Take a deep breath.
4. Place your mouth against that of the patient, ensuring that there is a good 'seal'.
5. Breathe out, inflating the patient's lungs. Watch the chest rise. This indicates a clear airway.
6. Remove your mouth and allow the patient's chest to fall.
7. Repeat three more times. Sometimes this will be sufficient to start the patient breathing.
8. Check that the heart is still beating.
9. If the patient is still not breathing but the heart is beating continue with the resuscitation at normal breathing rate (16–18 times per minute).
10. If the heart is not beating external compression must be given (see Section 8.4.3).

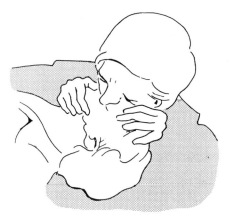

FIGURE 8.3 Mouth-to-mouth resuscitation. Reproduced from the *Approved Code of Practice for First Aid at Work* (HSE, 1985) by permission of the Health and Safety Executive.

If it is impossible to get air into the lungs there may be an obstruction to the airway or the patient's head is not in the correct position.

8.4.3 External chest compression (ECC)

The following procedure should be followed:

1. Kneel by the patient and find the correct position on the sternum: this is just below half way between the top at the neck and the bottom at the lower end of the chest.
2. Place the heel of one hand on the sternum in this position.
3. Place the other hand on top of the first and interlock the fingers.
4. Straighten the arms and rock forwards and backwards so that the body weight presses on the patient's chest. Do not jab or this may cause injury. The patient's chest should be compressed vertically about 40–50 mm.

To be effective, resuscitation and ECC should be given together. The rate should be 15 compressions to two inflations per minute. The pulse and breathing should be checked after one minute and then every three minutes. It should continue until either the patient responds or medical aid arrives. If the heart starts beating resuscitation should be continued until breathing is restored, when the patient should be placed in the recovery position.

8.4.4 Recovery position

The recovery position is shown in Fig. 8.2 and on the Guidance Chart in the first aid box:

1. Kneel at the side of the patient.
2. Turn the head towards you.
3. Push the nearest arm under the patient's back.
4. Pull the other arm over the patient's chest.
5. Cross the far leg over the nearest leg.
6. With one hand grasp the patient's clothing by the far hip (the waistband or belt usually ensures a tight grip).
7. Pull the patient over on to your knees, while protecting the head with your other hand.
8. Bend the upper leg forward.
9. Free the lower arm and extend it backwards.
10. Place the upper arm in a bent position forwards.
11. Tilt the patient's head back to ensure a clear airway.
12. Check the patient's breathing.

8.5 EXTERNAL BLEEDING

Only those types of bleeding that are associated with laboratory accidents are mentioned here. For other kinds the first aid manuals should be consulted.

Cuts from broken glass, microtome knives, scalpels and other 'sharps' and needlesticks make up the majority of such accidents. Animal bites and scratches also occur. Apart from action to arrest bleeding the possibility that the instrument might be contaminated with pathogens should be considered. Such injuries should be reported and medical advice obtained. Treatment or prophylaxis may be necessary to avoid a laboratory associated infection.

8.5.1 Treatment

1. Remove any loose foreign bodies that can be wiped or washed away. Do not attempt to remove embedded objects.
2. Apply firm pressure to the wound, directly if possible. If the wound is large squeeze it from either side.
3. Maintain pressure and elevate the bleeding part.
4. Apply a sterile dressing while maintaining pressure.
5. Maintain pressure for 15 minutes.
6. If bleeding continues and soaks through the dressing apply another dressing on top. Do not remove the first dressing.
7. Try to immobilize the injured part, e.g. place an arm in a sling.
8. If bleeding continues, reapply pressure, relieving it every 15 minutes until help arrives.

8.5.2 Severe bleeding

If direct pressure cannot be applied, pressure points may have to be used. The two common pressure points are (a) the brachial, between the muscles on the inside of the upper arm; (b) the femoral, in the fold of the groin.

As with direct pressure, pressure on these points should be maintained only for 15 minutes at a time.

8.5.3 Minor cuts

Minor cuts should be washed well under running water, dried and covered with sterile adhesive dressings. They should not be squeezed to encourage bleeding. Infection risks should be considered.

8.6 BURNS AND SCALDS

Burns and scalds are very common in laboratories and may be caused by dry heat (bunsens, hot air ovens, hot glass, outer cases of heated apparatus); electricity; corrosive chemicals;

cryogenic materials (liquefied gases, solid carbon dioxide; moist heat (boiling water, steam, culture media); contact with moving machinery (friction burns).

The skin is a good insulator but when it is burned or scalded the action of the heat will continue into the deeper layers, even after the cause has been removed. Shock and loss of body fluids may result. All except very minor burns require medical attention.

8.6.1 Treatment of burns and scalds

1. Remove the cause. Take care not to be affected, e.g. by electricity.
2. Cool the affected part with cold water, by immersion if possible, for at least ten minutes.
3. Remove rings, shoes, tight articles before swelling begins.
4. Do not remove clothing unless it can be 'floated off' during the water treatment.
5. When pain has lessened cease water treatment.
6. Cover the area completely with a dry sterile dressing or cloth. Do not use adhesive dressings on any burns. Bandage loosely to avoid the bursting of any blisters.
7. If the patient is conscious give fluids (not alcohol) in frequent small quantities to treat help replace fluid loss.
8. Never apply ointments, greases etc.

If the outside of the neck is affected (or chemicals have been swallowed) there may be swelling of the throat and breathing problems. If the patient is conscious give sips of cold water. If breathing stops apply artificial respiration. If the patient is breathing but unconcious place him or her in the recovery position.

8.6.2 Chemicals on the skin and clothing

Chemicals may cause burns. Treatment is as follows:

1. Drench the affected area with water using a shower or a hose.

2. Remove clothing during drenching.
3. Wash skin thoroughly with more water.

NB Observe local rules for skin contact with certain chemicals.

8.7 EYE INJURIES

The majority of eye injuries in laboratories are preventable (see Section 1.7.4). Medical advice should be sought in all cases of eye injury - chemical splashes, foreign bodies etc.

8.7.1 First aid for chemical splashes in the eye

Acids coagulate protein and form a shield which prevents further injury but caustic substances do not and continue to burn into the eye.

1. If contact lenses are worn remove them.
2. Incline the head so that the affected eye is below the uninjured eye and injurious material is not washed from one to the other.
3. Make sure that the injured eye is fully open so that it can be irrigated.
4. Flush the eye with water from a tap or eye-wash bottle for 10–20 minutes, longer if the chemical was alkaline.
5. Apply an eye pad and get medical assistance.

8.7.2 Foreign bodies

Remove only foreign bodies that can be washed out with water as above or removed by gentle application of a swab or the corner of a wet handkerchief. Otherwise apply a pad and get the patient to a hospital as quickly as possible.

8.7.3 Ultraviolet (uv) radiation

Exposure to uv, especially in the biocidal range, can cause headaches, conjunctivitis or temporary blindness. Bathe the eyes in water to alleviate irritation and remove the patient to a hospital.

8.8 POISONS

Poisons may be ingested, absorbed through the skin or inhaled. Poisoning may be suspected if there is an unusual odour around the patient's mouth, his lips are discoloured, he complains of burning in the mouth and throat and if he is found unconscious in unusual circumstances, e.g. with opened bottles nearby.

8.8.1 First aid treatment

1. Get medical attention immediately. One person should do this while another renders first aid.
2. If the patient is conscious ask what substance was taken.
3. If unconscious but breathing place the patient in the recovery position (see Section 8.4.4).
4. If the patient is vomiting place him or her in the recovery position so that vomit is not inhaled.
5. If breathing has stopped apply resuscitation - but remember that the mouth-to-mouth method may be hazardous if poison has been ingested.
6. Do not induce vomiting. If the patient has swallowed a corrosive substance, alkali, petroleum product or bleach give small amounts of water or milk to relieve pain in the mouth and throat and to dilute the poison in the stomach.

The patient must be removed to hospital as quickly as possible. Information about the poison, and if available the poison itself, together with any vomit, should be sent with the patient.

8.9 ELECTRIC SHOCK

In the event of an electric shock:

1. Do not touch the victim if he or she is still in contact with the source.

2. Switch off the current. If the switch is not immediately accessible pull out the cable. Do not cut the cable.
3. Summon medical assistance.
4. If the patient has stopped breathing begin mouth-to-mouth resuscitation.
5. When breathing commences treat other injuries, e.g. burns.

8.10 PHYSIOLOGICAL SHOCK

Physiological shock may be induced by injury and requires attention. It may be recognized by:

1. patient looking pale and complaining of feeling cold;
2. cold and clammy skin;
3. shallow breathing;
4. nausea or vomiting.

8.10.1 Treatment of shock

1. Remove the immediate cause.
2. Send for medical assistance.
3. Keep the victim lying down.
4. Keep him or her warm.
5. If the patient is vomiting and no bones are broken place him or her in the recovery position. Otherwise turn the head to arch the neck. Elevate the legs if possible.
6. If the patient is conscious give fluids (tea, coffee, milk) unless otherwise indicated (e.g. in certain kinds of poisoning, see Section 8.8.1 and in suspected abdominal injuries).

8.11 OTHER CONDITIONS

Although the above information is directed towards accidents that might be peculiar to laboratories almost any other kind of accident may happen in and around them. Frequent study of the first aid handbooks (see References) is therefore recommended.

8.12 REFERENCES

HSE (1985) *Approved Code of Practice for First Aid at Work*. London: HMSO

St. John's Ambulance, British Red Cross and St. Andrew's Ambulance Association (1982). *First Aid Manual*. London: Dorling Kindersley.

Order of St. John (1982) *First Aid at Work*. London: O. St. J.

9 SAFETY AUDIT CHECK LIST

W.J. Gunthorpe and C.H. Collins

This list is not exhaustive and is intended as a guide to safety checks that should be carried out at regular intervals by safety officers, safety representatives and heads of departments.

9.1 THE PREMISES

1. Is the working space in each laboratory room adequate for the number of staff?
2. Are the floors of rooms and corridors covered with non-slip material which is in good repair and free from defects?
3. Are the rooms uncluttered and free from obstructions?
4. Are the rooms clean, especially in corners?
5. Are there any defects in the treads and risers of stairways and are all stairways fitted with handrails?
6. Are bench surfaces resistant to solvents and corrosive chemicals?
7. Is the furniture in good repair and free from obvious defects?
8. Are there hand basins, provided with hot and cold water, soap and paper towels, in each laboratory room and in specimen reception areas?
9. Are the toilet facilities clean and adequate for the numbers of staff (i.e. separate male and female lavatories, 1 per 15 staff members)?

10. Does each member of the staff have a private locker for street clothing and personal effects?
11. Are these lockers conveniently placed, near to the workplace?
12. Is there a proper rest room for staff?
13. Are showers or drenches provided for decontamination after chemical accidents?

9.2 SECURITY

1. Is the laboratory building locked when unoccupied?
2. Are individual rooms, e.g. where hazardous materials or expensive equipment are housed locked when unoccupied?
3. Are the key-holding arrangements such that unauthorized persons cannot easily gain access?

9.3 HEATING AND VENTILATION

1. Is there a comfortable working temperature at the beginning of the day?
2. Is it regulated according to external temperature or by the calendar?
3. Is there an uncomfortable solar gain during the summer months?
4. Is the temperature maintained during the winter months (min 16°C)?
5. Are adequate blinds fitted to windows that are exposed to full sunlight?
6. Is the ventilation adequate for the comfort of the staff - at least six air changes per hour - especially in internal rooms with mechanical ventilation?
7. Does the air normally flow from 'clean' to 'dirty' areas, i.e. from corridors to laboratories and not vice versa?
8. Does mechanical ventilation compromise extraction of contaminated air by safety cabinets and fume cupboards?

9.4 LIGHTING

1. Is the general illumination adequate (300–400 lux)?
2. Are there dark corners in rooms or corridors?
3. Is local lighting, e.g. at individual workbenches provided?

9.5 FIRE PRECAUTIONS

1. Are 'fire stations' provided with the correct appliances and extinguishers?
2. Does each laboratory room have an appropriate fire extinguisher and fire blanket?
3. When were the fire extinguishers last checked?
4. Are fire exits free from obstructions?
5. Are fire exits clearly signed?
6. Where is the fire assembly point?
7. When was the last fire drill?

9.6 SIGNS AND NOTICES

1. Are direction signs adequate, sensibly worded and sensibly placed?
2. Are all warning and information signs (fire, hazard, first aid etc) of internationally agreed size, colour and wording.
3. Are notices diplayed on the doors and external windows of rooms where hazardous chemicals, gas cylinders, radioactive materials and high risk micro-organisms are handled or stored?

9.7 SERVICES

1. Are there enough sinks, gas points and electrical socket outlets in each room and at each bench?
2. Are there long and/or trailing cables or pipes that might trip people?
3. Is there a direct and effective system for reporting service faults?

4. Are such faults corrected within a reasonable time for the laboratory to function effectively?
5. Are service installations, cables, pipes, fuses, plugs etc. inspected and maintained on a regular basis?

9.8 STAFF

1. Is there (a) a safety officer (b) a safety representative (c) a safety committee?
2. Have all members of the staff been given copies of local and national codes of practice?
3. Have they been given induction training and made familiar with laboratory geography and customs?
4. Is the dress and/or behaviour of any member of the staff such as might be a hazard to him or her or his or her colleagues?

9.9 PROTECTIVE CLOTHING

1. Do the overalls/gowns/coats provided conform to the requirements of the Howie Code (e.g. Dowsett–Heggie or similar pattern)?
2. Are plastic aprons and gloves also provided for work with (a) micro-organisms as specified by the ACDP, (b) carcinogens, (c) radioactive substances?
3. Are rubber aprons and disposable gloves provided for (a) work with hazardous chemicals, including strong disinfectants, (b) clearing up after spillages?
4. Are heat-resistant gloves provided for (a) unloading sterilizers, (b) cryogenic materials?
5. Are goggles, vizors and/or safety spectacles provided for specified procedures?
6. Are vizors provided (and worn) for autoclave unloading?

9.10 HEALTH OF STAFF

1. Is there an occupational health service?
2. Do all members of the staff know where it is?

3. Are members of clerical and reception staff instructed in the potential hazards of handling specimens?
4. Are women of child-bearing age warned of the consequences of work with certain agents (e.g. rubella, cytomegalovirus)?
5. Are they told that if they suspect that they are pregnant they should inform the occupational health doctor or a medically qualified member of the laboratory staff so that alternative arrangements may be made for them?
6. Is there an inoculation/immunization programme relevant to the infectious material that is handled?
7. Are chest X-rays arranged for staff who are exposed to tubercle bacilli?
8. Are injuries and incidents that might affect the health of staff always recorded in the official accident/incident record books?

9.11 FIRST AID

1. Are first aid boxes provided at strategic points?
2. Do their contents conform to the HSE requirements and are they refilled at regular intervals?
3. Are first-aiders known to other staff?
4. In laboratories where there is no statutory requirement for first-aiders, is anyone trained to give first aid to victims of accidents peculiar to laboratories, e.g. contact with strong acids, corrosive chemicals, accidental ingestion of poisons?
5. Are there prominent notices giving emergency telephone numbers?

9.12 UNSAFE PRACTICES

1. Are smoking, eating and drinking permitted in laboratory rooms?
2. Is mouth pipetting permitted?
3. Are staff permitted to visit areas other than laboratories while wearing laboratory protective clothing?
4. Are food and drink stored in laboratory refrigerators?

9.13 CHEMICALS

1. Are all chemicals properly labelled with names and appropriate warning signs?
2. Are wall charts of warning and caution signs prominently displayed in rooms where chemicals are used or stored?
3. Is equipment provided for dealing with spillages?
4. Are staff trained to deal with spillages?
5. What quantities of flammables are kept in laboratory rooms?
6. Are flammables stored in unmodified refrigerators?
7. Are they stored in laboratories in officially approved fire-resistant cabinets?
8. Are drip trays provided for strong acids and corrosives?
9. Are carriers provided for large bottles?
10. Are compressed gas cylinders secured so that they cannot fall?
11. Are trollies provided for compressed gas cylinders?
12. Are compressed gas cylinders other than those in current use stored in the laboratory rooms?
13. Are hydrogen and LPG cylinders near to bunsens or other flames?
14. Are stock chemical and flammable stores conveniently placed?
15. Is there a sill to prevent spilled chemicals leaking under the door of chemical stores?
16. Are chemicals stored with proper regard to incompatibilities and proximity to heat and light?
17. Are fume cupboards used for storing chemicals?

9.14 ISOTOPE/RADIOACTIVE MATERIALS

1. Do the rooms conform to national radiation protection standards?
2. Do the staff know where to contact the radiation protection officer and radiation safety officer?
3. Are there written instructions about dealing with spillages?
4. Are isotopes stored safely?
5. Are there monitoring facilities for premises and staff?

6. Are there regular surveys of radiation levels?
7. Are suitable records kept?
8. Does disposal practice conform to national and local requirements?
9. Is the Radioactive Substances Licence clearly displayed?

9.15 EQUIPMENT

1. Does electrical equipment conform to the ESCHLE standards?
2. Is electrical and automated equipment regularly inspected and maintained under contract?
3. Are cables and fuses checked regularly?
4. Are autoclaves and other pressure vessels given independent checks (e.g. by insurance engineers)?
5. Are desiccators and vacuum equipment guarded against implosions?
6. Are microtome knives and other sharps stored safely?
7. Are the air flows in and around safety cabinets and fume cupboards checked by independent professionals?
8. Are the filters of safety and clean air cabinets tested regularly according to manufacturers' and official standards?
9. Are pipetting devices provided and are they appropriate to the work?
10. Are hypodermic needles and syringes used instead of pipettes and pipetting devices?

9.16 GLASS

1. Are there suitable receptacles for the safe discard of broken glass?
2. Are safety instructions given to staff who cut and manipulate glass tubing?
3. Is glassware regularly inspected for scratches and chips?
4. Is such glassware always discarded?
5. Are plastic pipettes and other equipment used instead of glass wherever possible?

9.17 INFECTIOUS MATERIAL

1. Are specimens submitted in a safe manner?
2. Are they unpacked with due regard to leakage and contamination?
3. Are gloves worn for unpacking specimens?
4. Do all members of the staff know what to do if infectious material is spilled or aerosols are known to have been released?
5. Are Danger of Infection (High Risk) specimens so labelled?
6. Are staff who work in Containment (Level 3) laboratories trained in the use and limitations of safety cabinets?
7. Are sealed buckets provided for centrifuges?
8. Is automated equipment tested for dispersal of aerosols and infectious particles.
9. Is infectious material allowed to accumulate on benches?
10. Is discarded material allowed to stand in containers on the floor overnight?
11. Are disinfectants used correctly, at the appropriate dilution and replaced daily?
12. Are benches and worktops regularly cleaned and disinfected?
13. Are satisfactory disposal containers (e.g. plastic bags in containers with leak-proof bottoms and sides) provided?
14. Are autoclaves loaded correctly?
15. Are autoclave performances checked (a) continually by instrumentation (b) regularly by biological tests?
16. Is infectious material carried to incinerators by staff other than laboratory employees?
17. Are the provisions of the Howie Code as amended by the HSAC and ACDP and other appropriate codes and legislation generally observed?

APPENDIX 1

Acts of Parliament and Regulations cited in the text

Carcinogenic Substances Regulations 1967 amended 1973.
Control of Pollution Act 1974.
Control of Pollution (Special Waste) Regulations 1980.
Control of Substances Harmful to Health 1987 (COSHH).
Deposit of Poisons Regulations 1972.
Disabled Persons Employment Acts 1944, 1958.
Health and Safety at Work etc. Act 1974.
Health and Safety (First Aid) Regulations 1981.
Ionizing Radiation Regulations 1985.
Poisons Act 1972.
Poisons Rules 1978.
Poisons List Order 1978.
Public Health Act 1971.
Radioactive Substances Acts 1948, 1960.
Reporting of Injuries, Diseases and Dangerous Occurrences
    Regulations 1985 (RIDDOR).
Safety Signs Regulations 1980.

INDEX